Lipids and Lipoproteins

Proceedings of a Scientific Symposium of the
International Federation of Margarine Associations
Brussels, May 17–18, 1979

Guest Editor: *M. Fondu,* Brussels

76 figures, 22 tables, 1980

S. Karger · Basel · München · Paris · London · New York · Sydney

Reprint from 'Nutrition and Metabolism',
Vol. 24, Supplement 1, 1980

Scientific Committee:

J. Lederer, Louvain
H. Greten, Heidelberg
A.J. Vergroesen, Vlaardingen
J.P. Wolff, Paris
M. Fondu, Brussels

S. Karger AG, P.O. Box, CH–4009 Basel (Switzerland)
Printed in Switzerland by Werner Druck AG, Basel
ISBN 3–8055–1266–X

Contents

ContentsIV

Nutr. Metab. *24* (Suppl. 1): 1–2 (1980)

Structure and Metabolism of Plasma Lipoproteins – An overview

W. Brown

Department of Medicine, Mount Sinai Medical Center, New York, N.Y.

Chylomicrons and very-low-density lipoproteins (VLDL) are the circulating particles responsible for transport of triglycerides from intestine and liver, respectively. Cholesterol ester exists with the triglyceride as a droplet surrounded by a monolayer of phospholipid-free cholesterol and specific proteins.

These apolipoprotein now number nine.

Apolipoprotein B (apo-B) is an essential component of both chylomicrons and VLDL. Other apolipoproteins are removed during triglyceride hydrolysis by lipoprotein lipase in the peripheral tissues (e.g. muscle and adipose tissue) but apo-B remains with the residual lipoprotein particles in the bloodstream. This 'remnant lipoprotein' may be taken up by the liver or converted by poorly understood processes to low-density lipoprotein (LDL). The latter is triglyceride-poor but cholesterol-rich. Its long half life in plasma (3 days) results in a large plasma pool of cholesterol in humans and is strongly associated with cardiovascular mortality.

Studies in our laboratory have examined the relation between dietary changes and the turnover of apo-B in VLDL and LDL. The first series of studies contrasted the effects of a diet containing 80% carbohydrate and 20% protein to a control diet of 40% carbohydrate, 40% fat and 20% protein. In 6 subjects with mild hypertriglyceridemia (type IV), triglycerides from 150 to 350 mg/dl the VLDL was isolated and labeled with ^{125}I (50–75 μCi) and triglycerides were labeled by intravenous injection of H-glycerol (300 μCi). The high-carbohydrate diet increased the production rate of triglyceride by 10% in VLDL but did not change the VLDL apo-B production. Thus the output of the major lipid of this lipoprotein is not directly tied to the synthesis and secretion of its major structural protein. A second finding was an increase in the removal rate of apo-B as a component of VLDL before its final conversion to LDL apo-B. Thus the apparent synthesis of LDL apo-B was reduced during a period of increased triglyceride synthesis.

The feeding of a diet high in polyunsaturated fats (P/S–1/1) from safflower oil was compared to a diet containing lard (P/S–0.2) ^{125}I-LDL was used to measure the rates of synthesis and removal of this lipoprotein. After 3 weeks of equilibration on each diet, the labeled lipoprotein was injected intravenously and monitored for an additional 3 weeks. 7 normal subjects and 8 individuals with familial hypercholesterolemia (heterozygotes) were studied on both diets. In both groups, the major effect was a reduction in the synthesis of LDL. The synthetic rate in normals fell by approximately 15% while that in the hypercholesterolemic subjects fell by 20–50%. In normals, there was also some increase in the removal rate but in the hypercholesterolemics no consistent change was seen in clearance. Thus, large reductions in saturated fat with replacement by polyunsaturated fat have a profound effect on the production of the major atherogenic lipoprotein in man.

W. Brown, MD, Mount Sinai Medical Center, Department of Medicine,
1, Gustave L. Levy Place, New York, NY 10029 (USA)

Nutr. Metab. *24* (Suppl. 1): 3–11 (1980)

Intestinal Lipoprotein Formation

Robert M. Glickman

Columbia University, New York, N.Y.

Key Words. Intestinal lipoproteins · Chylomicrons · Triglyceride · Lipid absorption · Apoproteins

Abstract. The average western diet contains approximately 40% of total calories as dietary fat or approximately 100 g of fat. The efficiency of the entire process of fat absorption can be judged by the fact that under normal conditions less than 5% of ingested fat is recovered in the stool. In the past several years, new concepts have greatly added to our understanding of the process by which dietary fat is digested, absorbed and processed in the intestinal epithelial cell for delivery to the body via the intestinal lymph and the portal venous system. These newer concepts include an understanding of the physical chemistry of lipids, the physiology of bile salts and the formation and metabolisms of lipoprotein all directly influencing the process of fat absorption. The present discussion will emphasize the formation of lipoproteins within the intestinal mucosa. New information suggests that the small intestinal mucosa is a quantitatively important source of lipoprotein constituents for systemic lipoproteins. This is hardly surprising when one considers the large quantities of lipid traversing the intestinal mucosa each day which must exit in the form of lipoproteins.

Mucosal Phase of Fat Absorption

Successful micellarization of lipids within the intestinal lumen permits these lipid products to diffuse to the surface of the intestinal epithelium and make intimate contact with the microvillus membrane. This is particularly important since it has been shown experimentally that surrounding the surface of the intestine is an unstirred water layer which functionally may pose a significant barrier to the diffusion of hydrophobic molecules such as lipids. This relatively immobile aqueous layer is more easily penetrated by the

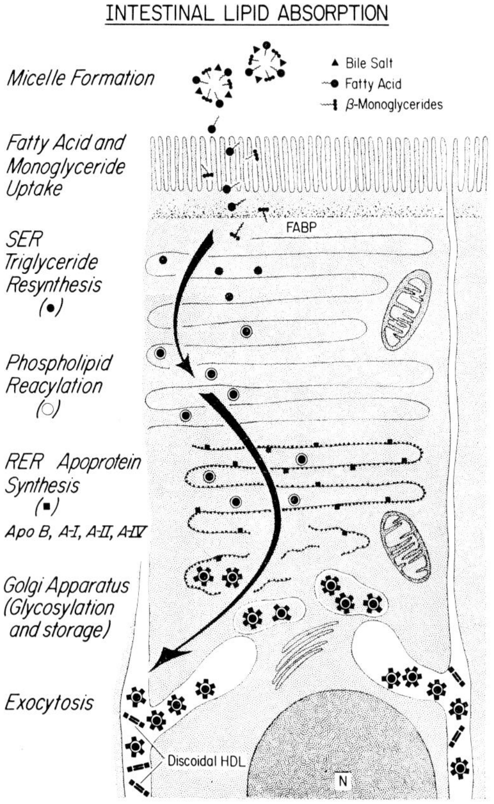

Fig. 1. Schematic representation of intestinal epithelial cell during fat absorption. FABP = Fatty-acid-binding protein; SER = Smooth endoplasmic reticulum; RER = rough endoplasmic reticulum; HDL = high-density lipoprotein; N = nucleus.

micellar complex thus increasing the efficiency of lipid uptake into the intestinal mucosal cell.

The uptake of lipids such as fatty acids and monoglycerides across the microvillus membrane is a passive process and results from the solubility of the lipid moieties within the lipid rich surface membrane of the epithelial cell. Recently, a low-molecular-weight cytosolic protein, fatty-acid-binding protein has been isolated from the intestine [1]. This protein avidly binds fatty acids and appears to function as an intracellular transport protein for long-chain fatty acids. Under experimental conditions where fatty-acid binding to this protein is inhibited, less fatty acid is available for triglyceride resynthesis. Thus it appears that fatty-acid-binding protein may serve a transport function within the intestinal epithelium and directs intracellular fatty acids to the smooth endoplastic reticulum, the site of triglyceride resynthesis. A schematic representation of events within the cell is shown in figure 1. Triglyceride resynthesis reduces the effective concentration of free fatty acids within the cell and maintains an effective concentration gradient for the continued passive uptake to take place. In addition, the storage of fatty acids as the more inert triglycerides while awaiting transport from the intestine may spare the cell the potential injurious effects of high intracellular free fatty acid concentration.

The enzymes for triglyceride resynthesis have been localized biochemically in the smooth endoplasmic reticulum in the apical portion of the intestinal epithelial cell beneath the microvillus membrane. This is corroborated morphologically in that the earliest time triglyceride can be visualized within the intestinal epithelial cell is within the profiles of the smooth endoplasmic reticulum [2]. Shortly after lipid absorption, the entire apical portion of the intestinal cell is filled with triglyceride droplets. Morphologically one can follow with time the movement of these triglyceride droplets through the profiles of the endoplasmic reticulum to the Golgi apparatus in the supranuclear portion of the cell. Here clusters of chylomicrons can be seen within the saccules of the Golgi apparatus. While few details are known concerning the actual mechanisms of lipoprotein egress from the cell, morphological studies show a migration of Golgi vesicles towards the laterobasal membrane where they appear to fuse with the membrane and discharge their contents into the intercellular space by reverse pinocytosis. There is experimental evidence to suggest that this directed intracellular movement may in part depend on intact microtubular function [3].

Characteristics of Intestinal Lipoproteins

Although the morphologic events of triglyceride transport have been well defined, less is known concerning the biochemical events of chylomicron formation. A great deal can be learned by analyzing the composition of chylomicrons obtained from the mesenteric lymph of laboratory animals [4]. Table I shows the chemical composition of rat intestinal lymph chylomicrons and the composition of chylomicrons obtained from the urine of 2 patients with chyluria due to long-standing filarial disease and documented

Table I. Characteristics of rat and human lymph chylomicrons

Chemical composition	Composition by weight	
	rat [4]	human [5]
Triglyceride, %	84	91
Phospholipid, %	13	7.5
Cholesterol, %	2	1.6
Protein, %	1	1.3
Density, g/ml	$<$1,006	$<$1,006
S_f value	$>$ 400	$>$ 400
Electrophoresis	origin	origin

communications between mesenteric and renal lymphatic systems studied in our laboratory [5].

It can be seen that triglyceride composes the major portion of chylomicron lipids which reflects the major physiological role of this particle in fat absorption. Phospholipid, although a smaller component, is important structurally. Together with free cholesterol and chylomicron protein it is arranged on the surface of this particle. Although the intestine can synthesize phospholipid *de novo* for the chylomicron surface, it appears that at least a portion of chylomicron phospholipid is derived from luminal sources (i. e. biliary lecithin) after reacylation of absorbed lysolecithin.

The protein content of the chylomicrons has been the object of considerable interest. Although quantitatively small (1% of chylomicron mass), it is now apparent that chylomicron apoproteins contain a characteristic complement of specific proteins [6, 7]. Figure 2 shows the apoprotein composition of rat and human lymph chylomicrons after delipidation and electrophoresis on SDS-polyacrylamide gels. It can be seen that there is a remarkable similarity in the protein patterns from these two species. Of particular importance to intestinal lipid transport is apoB. In addition, apoA-I, the major apoprotein of circulating high-density lipoproteins in most species is also an important chylomicron component and comprises 20% of chylomicron protein in man [5] and 40% in the rat [8]. This apoprotein is also an activator of lecithin: cholesterol acryltransferase, a plasma enzyme responsible for cholesterol esterification. ApoA-IV is a newly described chylomicron apoprotein in man [9, 10] and is analogous to a similar apoprotein in the rat [11]. Its metabolic importance remains to be determined. In addition, a group of small-molecular-weight apoproteins,

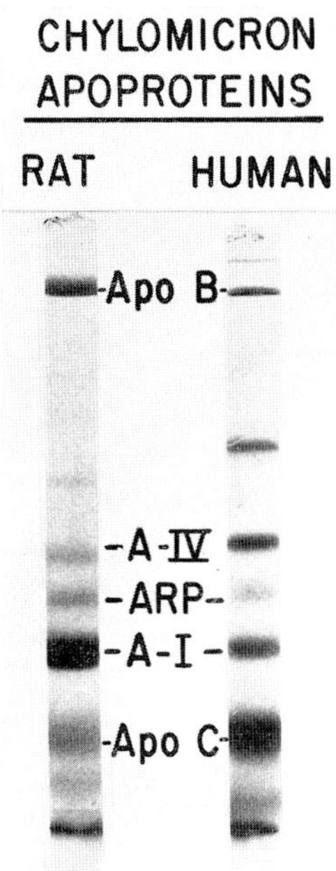

Fig. 2. Apoprotein composition of rat mesenteric lymph and human chylomicrons (isolated from chyluria). The apoproteins are separated on SDS-polyacrylamide gels. ARP = argenine rich protein.

the C apoproteins, is also found on lymph chylomicrons [6]. One of this group, apoC-II, is an activator of lipoprotein lipase and hence is of extreme importance in the catabolism of chylomicrons after secretion [12, 13]. A recent report [14] described a man with severe hypertriglyceridemia and a complete deficiency of this lipoprotein lipase activator, apoC-II, resulting in a marked impairment of chylomicron lipolysis. After partial replacement of apoC-II by transfusion of plasma from a normal subject the patient's triglyceride fell, within 1 day, from 1,000 to 250 mg/dl.

Since lymph is an ultrafiltrate of plasma and apoproteins have been shown to exchange among plasma lipoproteins, one cannot be certain that a given apoprotein, despite being associated with lymph chylomicrons, originates in the intestine. It is therefore necessary to determine which of these apoproteins is actively synthesized by the intestinal mucosal cell during chylomicron formation. Experimentally, one approach to this problem is to observe the incorporation of radioactive amino acids into the various chylomicron apoproteins of lymph during lipid absorption in the rat [6, 8]. Only three chylomicron apoproteins appear to be synthesized during chylomicron formation – apoB, apoA-IV and apoA-I. It appears that the other chylomicron apoproteins are acquired by the particle after secretion by the intestinal cell via transfer from other lipoproteins.

Importance of Intestinal Apoprotein Synthesis

The demonstration of apoB in intestinal epithelial cells and its role during fat absorption has been demonstrated using fluorescent antibody techniques [15]. A progressive, rapid increase in apoB fluorescence, which gradually fills the entire apex of the cell occurs early in fat absorption. The distribution of fluorescence in the apical and supranuclear portions of the cell suggests that a pool of the apoprotein is present within the cell and, with active lipid absorption, there is a progressive increase in apoB fluorescence consistent with the known increase in synthesis of this apoprotein during chylomicron formation [16].

Although quantitatively a small proportion of chylomicron mass, apoprotein synthesis appears to be extremely critical for triglyceride transport through the intestinal mucosa. It has recently been shown that the intestinal mucosa of patients with abetalipoproteinemia [17], a rare hereditary disorder associated with a total inability to form triglyceride-rich lipoprotein [18], despite being engorged with triglyceride, completely lacks immunoreactive apoB [17] and this apoprotein is also absent in the plasma. It appears that other apoproteins are normal. Thus, this rare disease has clearly shown that apoB synthesis is an obligatory step in chylomicron formation. Although there are no human disease states with an inability to synthesize apoA-I or apoA-IV, experimental studies in animals suggest that these apoproteins are not absolutely required for chylomicron formation to proceed. Similar biopsies showing engorgement of the mucosa with triglyceride have been seen in human protein calorie malnutrition [19] as well as experimental conditions in animals with impaired protein synthesis [8]. Thus intact protein synthesis appears to be necessary for chylomicron formation to proceed.

It has been shown in experimental animals [7, 15, 20] and man [18, 21] that intestinal mucosa in the fasting state contains a small pool of chylomicron apoproteins. This pool becomes rapidly depleted during the initial stages of chylomicron formation and new apoprotein synthesis is required to permit continued chylomicron formation to proceed. Experimentally, once this pool is depleted and protein synthesis is impaired triglyceride accumulates within the cell.

Contribution of the Intestine to Systemic Lipoprotein Metabolism

There is mounting evidence that the intestine is a major synthetic source of apoprotein constituents of important plasma lipoproteins. As noted above, apoproteins such as apoB, apoA-I and apoA-IV are actively synthesized by the intestine during triglyceride absorption. The intestine of the rat also secretes discoidal high-density lipoproteins [22], an additional source of apoA-I for plasma. Since these lipoproteins of intestinal origin enter the systemic circulation, they directly contribute to the levels of these apoproteins in plasma. This is especially true for apoA-I which has been shown to leave the chylomicron surface after secretion and eventuate in plasma high-density lipoproteins [23, 24]. Estimates in man indicate that as much as 30% of the total daily synthesis of this apoprotein may originate in the intestine and thus directly influence plasma high-density lipoprotein metabolism [5]. Further research is required to determine factors which modulate the intestinal synthesis of such apoproteins. Sufficient data are already available to indicate that dietary influences and their effect on the quantitative and qualitative aspects of intestinal lipoprotein formation will have important consequences for systemic lipoprotein metabolism.

References

1   Ockner, R.K. and Manning, J.A.: Fatty acid binding proteins. Role in esterification of absorbed long chain fat in rat intestine. J. Clin. Invest. *58:* 632 (1976).
2   Cardell, R.R., Jr.; Badenhausen, S., and Porter, K.P.: Intestinal triglyceride absorption in the rat. An electron microscopical study. J. Cell Biol. *34:* 125 (1967).
3   Glickman, R.M.; Perrotto, J.L., and Kirsch, K.: Intestinal lipoprotein formation: effect of colchicine. Gastroenterology *70:* 347 (1976)
4   Glickman, R.M.: Chylomicron formation by the intestine; in Rommel and Goelell, Lipid absorption: biochemical and clinical aspects, Ed. K. Rommel, Ulm (1975)
5   Green, P.H.R.; Glickman, R.M.; Saudek, C.D.; Blum, C.B., and Tall, A.R.: Intestinal lipoprotein secretion in chyluric man. J. clin. Invest. *64:* 233 (1979).
6   Imaizumi, I.; Fainaru, M., and Havel, R.J.: Composition of proteins of mesenteric lymph chylomicrons in the rat and alterations upon exposure of chylomicrons to blood serum and serum proteins. J. Lipid Res. *19:* 212 (1978).
7   Glickman, R.M. and Green, P.H.R.: The intestine as a source of apolipoprotein A-I. Proc. natn. Acad. Sci. USA *74:* 2569 (1977).

8 Glickman, R.M. and Kirsch, K.: Lymph chylomicron formation during the inhibition of protein synthesis. J. Clin. Invest. *52:* 2910 (1973).

9 Green, P.H.R.; Glickman, R.M.; Riley, J.W., and Quinet, E.: Human apolipoprotein A-IV intestinal origin. Gastroenterology *76:* 1143 (1979).

10 Weisgraber, K.H.; Bersot, T.P., and Mahley, R.W.: Isolation and characterization of an apoprotein from the $d < 1.006$ lipoproteins of human and canine lymph homologous with the rat A-IV apoprotein. Biochem. biophys. Res. Commun. *85:* 287 (1978).

11 Wu, A.L. and Windmueller, H.G.: Identification of circulating apolipoproteins synthesized by rat small intestine *in vivo.* J. biol. Chem. *253:* 2525 (1978).

12 Ganesan, D.; Bradfort, R.H.; Alaupovic, P., et al.: Differential activation of lipoprotein lipase from human post-heparin plasma, milk and tissue by polypeptides of human serum apolipoproteins C. FEBS Lett. *15:* 205 (1971).

13 Brown, W.V. and Baginsky, M.L.: Inhibition of lipoprotein lipase by an apoprotein of human very low density lipoprotein. Biochem. biophys. Res. Commun. *46:* 375 (1972).

14 Breckenridge, W.C.; Little, J.A.; Steiner, G.; Chow, A., and Poapst, M.: Hypertriglyceridaemia associated with deficiency of apolipoprotein C-II. New Engl. J. Med. *298:* 1265 (1978).

15 Glickman, R.M.; Khorana, J., and Kilgore, A.: Localization of apolipoprotein B in intestinal epithelial cells. Science *193:* 1254 (1976).

16 Windmueller, H.G.; Herbert, P.N., and Levy, R.I.: Biosynthesis of lymph and plasma lipoprotein apoprotein by isolated perfused rat liver and intestine. J. Lipid Res. *14:* 215 (1973).

17 Glickman, R.M.; Green, P.H.R.; Lees, R.S.; Lux, S.E., and Kilgore, A.: Immunofluorescence studies of apolipoprotein B in intestinal mucosa. Absence in abetalipoproteinaemia. Gastroenterology *76:* 288 (1979).

18 Levy, R.I.; Fredrickson, D.S., and Laster, L.: The lipoproteins and lipid transport in abetalipoproteinaemia. J. Clin. Invest. *45:* 531 (1966).

19 Theron, J.J.; Wittmann, W., and Prinston, J.G.: The fine structure of the jejunum in kwashiorkor. Exp. molec. Path. *14:* 184 (1971).

20 Schonfeld, G.; Bell, E. and Alpers, D.H.: Intestinal apoproteins during fat absorption. J. clin. Invest. *78:* 1539 (1978).

21 Rachmilewitz, D.; Albers, J.J., and Saunders, D.R.: Apoprotein B in fasting and postprandial human jejunal mucosa. J. clin. Invest. *57:* 530 (1976).

22 Green, P.H.R.; Tall, A.R., and Glickman, R.M.: Rat intestine secretes discoid nascent high density lipoprotein. J. clin. Invest. *61:* 528 (1978).

23 Tall, A.R.; Green, P.H.R., and Glickman, R.M.: Metabolic fate of chylomicron phospholipids and apoproteins in the rat. Circulation *58:* 11–15 (1978).

24 Redgrave, T.G. and Small, D.M.: Transfer of surface components of chylomicrons to the high density lipoprotein fraction during chylomicron catabolism in the rat. Circulation *58:* 11–14 (1978).

R. Glickman, MD, Columbia University, College of Physicians and Surgeons, 630 West 168 Street, New York, NY 10032 (USA)

Discussion

Dr. Miller: We have recently reported the presence of discoidal particles in the HDL_2 density range of hepatic venous blood from normal human subjects. We would like to know whether these particles originate in the liver or in the intestine. Have you looked for the presence of nascent HDL in mesenteric venous blood from the rat intestine?

Dr. Glickman: We have not been able to demonstrate discoidal HDL in rat mesenteric venous blood or any increase in venous apoA-1 concentration in comparison with systemic venous blood.

Dr. Greten: Can you briefly comment upon the role of crypt cells as to intestinal apoprotein synthesis. Histochemically, apoproteins have also been identified in these cells which are also capable of synthesizing cholesterol.

Dr. Glickman: To date there are no quantitative data on the distribution of intestinal apoA-1 between villous and crypt cells. Both intestinal cell types are active in cholesterol synthesis and there is a strong possibility that apoproteins may be synthesized in intestinal cells not normally involved in triglyceride absorption, i.e. crypt cells. This possibility remains to be directly verified.

Dr. Hartman: There is an increase in A_1 of about 20 mg/dl blood after fat feeding. What is the time course of this observation?

Dr. Glickman: We have not carried out a detailed time course of the appearance of chylomicron apoA-1 in peripheral blood after fat feeding. In recently published studies we have made similar measurements in the urine of chyluric individuals. The sequence of apoA-1 increase directly paralleled the appearance of triglyceride in the urine.

Nutr. Metab. *24* (Suppl. 1): 12–18 (1980)

Uptake and Cellular Degradation of Low-Density Lipoprotein

G. R. Thompson

MRC Lipid Metabolism Unit, Hammersmith Hospital, London

Key Words. LDL · Fibroblast · Receptor · Catabolism · Familial hypercholesterolaemia

Abstract. *In vitro* data suggest that low-density lipoprotein (LDL) is bound to specific receptors located in pits on the surface of fibroblasts by a high-affinity process and subsequently undergoes catabolism in lysosomes. This binding seems to be mediated by ionic interaction between LDL and its receptor, the latter being totally or partially absent from the fibroblasts of patients with familial hypercholesterolaemia (FH). *In vivo* data suggest that LDL is also catabolised by a concentration-dependent, low-affinity pathway which is probably mainly located in the liver. LDL catabolism is reduced in FH and after saturated fat feeding, whereas polyunsaturated fat has the reverse effect. Hypocatabolism of LDL alters LDL composition, accelerates atherosclerosis and may lead to premature death from coronary heart disease.

The main purpose of this paper is to examine certain aspects of low-density lipoprotein (LDL) uptake and degradation. These include the mechanisms involved, the physiological sites at which these processes take place and some of the factors which influence LDL catabolism *in vivo*. Most of the emphasis will be on reviewing the work of others rather than discussing the author's own work in this important field.

Consideration of the composition of LDL reveals that almost half of the mass of an LDL particle is taken up by cholesterol, most of which is esterified. The rest consists mainly of phospholipids (notably lecithin and sphingomyelin) and apolipoprotein B (apoB), which are present in roughly equal proportions. The remaining 10% or less consists of triglyceride. In structural terms, LDL is organised in a pseudomicellar manner, with a surface layer of free cholesterol, phospholipid and apoB and an inner core of cholesterol ester and triglyceride.

LDL Uptake and Degradation *in vitro*

The numerous studies of *Goldstein and Brown,* which were carried out mainly on cultured skin fibroblasts, have considerably advanced our knowledge of LDL catabolism in both health and disease. In particular, they have helped elucidate the pathogenesis of familial hypercholesterolaemia (FH). This dominantly inherited disorder affects 0.2% of the population of Britain and North America and is expressed twice as severely in homozygotes as in heterozygotes. Both forms of the disease are associated with a marked predisposition to premature death from coronary heart disease.

In 1973 *Goldstein and Brown* [1] reported that fibroblasts cultured from the skin of a patient with homozygous FH had a higher level of HMG-CoA reductase activity than that present in normal fibroblasts, although the latter's activity increased after prolonged incubation in a lipoprotein-free medium. Addition of LDL to the medium suppressed the HGM-CoA reductase activity in normal fibroblasts to its previous low level but had no effect on the FH cells. These findings led *Brown and Goldstein* [2] to examine the ability of fibroblasts to bind labelled LDL. They found that normal fibroblasts bound LDL by means of a high-affinity, saturable mechanism which was not demonstrable in fibroblasts from FH homozygotes. They then went on to compare several other aspects of LDL and cholesterol metabolism in the two types of fibroblast. FH fibroblasts not only failed to bind LDL normally but also failed to take it up to the same extent as normal fibroblasts. Consequently, neither the apoB nor the cholesterol ester of LDL were degraded at the normal rate. Furthermore, as mentioned earlier, HMG-CoA reductase activity was not suppressed in FH cells when LDL was added to the incubation medium nor did they esterify free cholesterol derived from LDL catabolism at the same rate as normal fibroblasts.

On the basis of the above findings, *Brown et al.* [3] proposed the following scheme for LDL uptake and degradation in normal fibroblasts: an initial step involving high-affinity binding of LDL to the cell surface, followed by endocytosis and lysosomal hydrolysis of its apoB and cholesterol esters. This results in the liberation of free cholesterol which suppresses HMG-CoA reductase activity and thus inhibits cholesterol synthesis; any excess free cholesterol gets esterified to cholesterol oleate by acyl CoA: cholesterol acyl transferase. *Goldstein and Brown* [4] had earlier proposed that FH is due to an inherited deficiency of specific, cell-surface receptors for LDL, this deficiency being partial in heterozygotes and total in homozygotes. Although the rates of uptake, and thus of degradation, of LDL may be similar in heterozygotes and controls when expressed in absolute terms, catabolism is subnormal in heterozygotes when expressed as the fractional catabolic rate (FCR), i.e. relative to pool size [5].

Recently, *Orci et al.* [6] investigated the morphology of the LDL receptor. Their transmission electronmicrographs show ferritin-labelled LDL binding to shallow pits on the surface of normal fibroblasts and illustrate the manner in which LDL particles subsequently get incorporated into the cell within endocytic vesicles. Their scanning electron micrographs of fibroblasts processed by the freeze-fracture technique show that normal cells have shallow pits to which LDL particles adhere, whereas cells from a homozygote have similar pits but without any adherent LDL. This suggests that LDL receptors are normally located in such pits but that although the pits are there in FH they lack LDL receptors.

The mechanism of binding of LDL to receptors has recently been explored by *Mahley et al.* [7]. These workers have shown that a subfraction of HDL known as HDL_C has an even greater affinity for fibroblasts than LDL itself. HDL_C does not contain apoB but is rich in apoE. Now apoB and apoE have one thing in common, which is that they are both rich in arginine. *Innerarity and Mahley* [8] inactivated these arginyl residues by incubating LDL and HDL_C with cyclohexanedione and showed that this blocked their high-affinity uptake by fibroblasts. These authors proposed, therefore, that high-affinity binding of LDL involves some form of interaction between the positively changed arginyl residues of apoB and the LDL receptor. That this interaction is ionic is supported by the recent observation of *Filipovic et al.* [9] that pre-incubation of LDL with sialic-acid-containing gangliosides alters its surface charge and inhibits its high-affinity uptake by fibroblasts.

During their fibroblast studies, *Goldstein and Brown* [10] had shown that LDL uptake and degradation took place not only by a high-affinity mechanism but also via a low-affinity, concentration-dependent pathway, the latter being the sole mechanism for LDL binding and degradation in homozygote fibroblasts. Recent studies by *Shepherd et al.* [11] with cyclohexanedione-treated LDL suggests that almost 70% of LDL catabolism occurs via this low-affinity pathway *in vivo* in normal subjects and more than 80% in patients with heterozygous FH. The latter observation fits in with our own data on LDL turnover in FH patients in whom we had acutely reduced LDL levels by plasma exchange. The constancy of the FCR during the period when the LDL concentration was increasing after plasma exchange is consistent with catabolism occurring mainly via a nonsaturable, concentration-dependent pathway [12].

In vivo Sites of LDL Catabolism

It had long been thought that the liver was the major site but this concept received a rude shock when *Sniderman et al.* [13] showed that total hepatectomy in pigs increased the rate of LDL catabolism rather than the reverse, as had been expected. Subsequently, *Sniderman et al.* [14] redressed the balance somewhat when they obtained data in man which showed that LDL cholesterol was present in higher concentration in aortic blood than in hepatic venous blood, unlike VLDL and HDL. This suggested that the liver secreted VLDL and HDL cholesterol into the circulation but removed LDL cholesterol.

The possibility that the liver is an important site of LDL catabolism is further illustrated by the data of *Levy and Langer* [15], who found that administration of cholestyramine resulted in a marked increase in the FCR of LDL. This effect must have been an indirect one because cholestyramine is not absorbed and acts solely by preventing the reabsorption of bile acids from the gut. One possible explanation is that the liver meets the extra demand for cholesterol arising from the cholestyramine-induced stimulation of bile acid synthesis by catabolising LDL cholesterol at a faster rate.

Factors Influencing LDL Catabolism

Hormonal factors exert an important influence as illustrated by the stimulatory effect of oestrogens on the catabolism of LDL in the perfused rat liver [16]. Data obtained in humans by *Walton et al.* [17] suggest that *L*-thyroxine has an analogous effect.

Data on the effects of dietary fat on LDL turnover in man are somewhat scarce. However, in one small study [18] changing from a high to a normal P/S ratio diet resulted in an increase in plasma LDL which was presumably secondary to the 21% decrease in FCR that was observed since synthesis did not change. Unfortunately, the differing cholesterol content of the two diets confuses the interpretation of these results. To circumvent this problem, we [19] altered the fatty acid composition of LDL lipids without changing the diet. We achieved this by administering intravenous solutions containing egg lecithin. This phospholipid is less polyunsaturated than endogenous LDL lecithin but readily exchanges with the latter. This results in an increase in the oleic:linoleic acid ratio in LDL lecithin which is accompanied by a similar change in LDL cholesterol esters. We assessed the results of this manoeuvre on LDL turnover in 4 healthy subjects and found that it caused a 15% increase in plasma LDL. This was apparently due more to the 11% decrease in FCR than to the 4% increase in synthetic rate. These findings are by no means conclusive but they do provide some support for the concept that one of the mechanisms whereby polyunsaturated fats lower the serum cholesterol is by enhancing LDL catabolism. Conversely, saturated fat may raise serum cholesterol levels by inducing hypocatabolism of LDL.

It seems likely that hypocatabolism of LDL is responsible for some of the biochemical consequences of FH. Recently, *Jadhav and Thompson* [20] confirmed that LDL from homozygotes is richer in cholesterol and sphingomyelin and poorer in lecithin than normal LDL, with heterozygotes occupying an intermediate position. We found that these abnormalities became less marked for 1–2 days after patients had undergone a therapeutic plasma exchange, and then recurred. This suggests that newly formed LDL particles entering the circulation after a plasma exchange have a normal composition in FH but that they become abnormal as a consequence of the low FCR and the associated prolongation of LDL half-life. If so, this implies that as LDL ages it accumulates sphingomyelin and cholesterol and loses lecithin. The clinico-pathological consequences of hypocatabolism of LDL are exemplified by the extensive and premature atheroma which accompanies FH. This disease provides a stark illustration of the need to be able to take up and degrade LDL efficiently and the fatal results of inheriting a gene which prevents these processes from occurring normally.

References

1   Goldstein, J.L. and Brown, M.S.: Familial hypercholesterolemia: identification of a defect in the regulation of 3-hydroxy-3-methylglutaryl coenzyme A reductase activity associated with overproduction of cholesterol. Proc. natn. Acad. Sci. USA *70:* 2804 (1973).
2   Brown, M.S. and Goldstein, J.L.: Familial hypercholesterolemia: defective binding of lipoproteins to cultured fibroblasts associated with impaired regulation of 3-hydroxy-3-methylglutaryl coenzyme A reductase activity. Proc. natn. Acad. Sci. USA *71:* 788 (1974).
3   Brown, M.S.; Luskey, K.; Bohmfalk, H.A.; Helgeson, J., and Goldstein, J.L.: Role of the LDL receptor in the regulation of cholesterol and lipoprotein metabolism; in Greten, Lipoprotein metabolism, pp. 82–89 (Springer, Berlin 1976).
4   Goldstein, J.L. and Brown, M.S.: Familial hypercholesterolemia. A genetic regulatory defect in cholesterol metabolism. Am. J. Med. *58:* 147 (1975).
5   Langer, T.; Strober, W., and Levy, R.I.: The metabolism of low density lipoprotein in familial type II hyperlipoproteinemia. J. clin. Invest. *51:* 1528 (1972).

6 Orci, L.; Carpentier, J.L.; Perrelet, A.; Anderson, R.G.W.; Goldstein, J.L., and Brown, M.S.: Occurrence of low density lipoprotein receptors within large pits on the surface of human fibroblasts as demonstrated by freeze-etching. Expl Cell Res. *113:* 1 (1978).

7 Mahley, R.W.; Innerarity, T.L.; Pitas, R.E.; Weisgraber, K.H.; Brown, J.H., and Gross, E.: Inhibition of lipoprotein binding to cell surface receptors of fibroblasts following selective modification of arginyl residues in arginine-rich and B apoproteins. J. biol. Chem. *252:* 7279 (1977).

8 Innerarity, T.L. and Mahley, R.W.: Enhanced binding by cultured human fibroblasts of apo-E-containing lipoproteins as compared with low density lipoproteins. Biochemistry, N.Y. *17:* 1440 (1978).

9 Filipovic, I.; Schwarzmann, G.; Mraz, W.; Wiegandt, H., and Buddecke, E.: Sialic acid content of low density lipoproteins controls their binding and uptake by cultured cells. Eur. J. Biochem. *93:* 51 (1979).

10 Goldstein, J.L. and Brown, M.S.: Binding and degradation of low density lipoproteins by cultured human fibroblasts. J. biol. Chem. *249:* 5153 (1974).

11 Shepherd, J.; Bicker, S., and Packard, C.T.: Receptor independent low density lipoprotein catabolism in man. Eur. J. clin. Invest. *9:* 11–32 (1979).

12 Thompson, G.R.; Spinks, T.; Ranicar, A.A., and Myant, N.B.: Non-steady-state studies of low density lipoprotein turnover in familial hypercholesterolaemia. Clin. Sci. mol. Med. *52:* 361 (1977).

13 Sniderman, A.D.; Carew, T.E.; Chandler, T.G., and Steinberg, D.: Paradoxical increase in rate of catabolism of low density lipoproteins after hepatectomy. Science *183:* 526 (1974).

14 Sniderman, A.; Thomas, D.; Marpole, D., and Teng, B.: Low density lipoprotein. A metabolic pathway for return of cholesterol to the splanchnic bed. J. clin, Invest. *61:* 867 (1978).

15 Levy, R.I. and Langer, T.: Hypolipidemic drugs and lipoprotein metabolism. Adv. exp. Med. Biol. *21:* 7 (1972).

16 Hay, R.V.; Pottenger, L.A.; Reingold, A.L.; Getz, G.S., and Wissler, R.W.: Degradation of I[125]-labelled serum low density lipoprotein in normal and estrogen-treated male rats. Biochem. biophys. Res. Commun. *44:* 1471 (1971).

17 Walton, K.W.; Scott, P.J.; Dykes, P.W., and Davies, J.W.L.: The significance of alterations in serum lipids in thyroid dysfunction. II. Clin. Sci. *29:* 217 (1965).

18 Levy, R.I. and Langer, T.: Lipoprotein metabolism; cited by Eisenberg, S. and Levy, R.I. in Adv. Lipid Res. *13:* 1 (1975).

19 Thompson, G.R.; Jadhav, A.; Nava, M., and Gotto, A.M.: Effect of intravenous phospholipid on low density lipoprotein turnover in man. Eur. J. clin. Invest. *6:* 241 (1976).

20 Jadhav, A.V. and Thompson, G.R.: Reversible abnormalities of low density lipoprotein composition in familial hypercholesterolaemia. Eur. J. clin. Invest. *9:* 63 (1979).

G. Thompson, MD, FRCP, MRC Lipid Metabolism Unit, Hammersmith Hospital, Ducane Road, London W12 OHS (UK)

Discussion

Prof. Verdonk: Most of the time, maturity-onset diabetics are obese. According to my experience, they need caloric restriction which causes their blood lipids to decrease, and their HDL cholesterol to increase. The clinical and biochemical approach for therapy in nature obese diabetics is to restrict calories; hypercholesterolaemia and hyperglyceridaemia will then disappear almost completely.

Dr. Thompson: I think that angiographic studies in patients with coronary heart disease have shown that if a patient has a hyperlipidaemia, in particular hypercholesterol-semia as his prime risk factor, he tends to present with his coronary heart disease at the younger age and with a more severe type of disease then when the latter is associated mainly with smoking or, maybe, hypertension. I wonder whether Dr. *Epstein* would like to comment on the likelihood that if diet is modifying coronary heart disease, he would expect to find the fall off in mortality most marked among the younger age group.

Dr. Epstein: This is exactly the situation in Finland, where cholesterol levels are spectacularly high. Children in Finland have average cholesterol levels around 200 in comparison with levels around 170 in other parts of Europe. I would say that the Finnish experience, and perhaps also the British one, where there is perhaps a trace of a declining trend in younger age groups, very much supports, on a population level, the results of angiographic studies.

Nutr. Metab. *24* (Suppl. 1): 19–25 (1980)

Role of Low-Density and High-Density Lipoproteins in Atherogenesis

G. Assmann and H. Schriewer

Zentrallaboratorium der Medizinischen Einrichtungen der Universität Münster, Münster

Key Words. Low-Density Lipoproteins · High-Density Lipoproteins · Atherogenesis · Tangier disease

Abstract. Among the cholesterol-carrying lipoproteins, low-density lipoproteins (LDL) have been associated with coronary heart disease as a risk factor while high-density lipoproteins (HDL) appear to protect against coronary heart disease. According to studies with cells in tissue culture, control mechanisms of receptor-mediated LDL uptake are important in maintaining the cholesterol balance within the arterial cells. HDL may be a vehicle for transporting cholesterol from peripheral cells to the liver. Recent results, derived from studies of patients affected with Tangier disease (absence of HDL in plasma), favor the hypothesis that HDL precursors (e.g. surface remnants of chylomicrons) may be more potent in cholesterol uptake than mature HDL.

In earlier epidemiologic studies, hypercholesterolemia has been associated with coronary heart disease as a risk factor [1–7]. These former studies almost exclusively focused on the elevation of total serum cholesterol as a good predictor for the risk of this disease. Recent studies, however, show that the evaluation of coronary risk based simply on total serum cholesterol levels was too restrictive [8]. According to the Framingham study, it could be shown that the risk was not confined to those individuals with extreme hypercholesterolemia, but also included individuals in the normal to border-line range of 200–300 mg/dl. Thus, young victims of coronary infarcts under the age of 50 years in this study had an average serum cholesterol of only 244 mg/dl.

Low-Density Lipoproteins

Since total serum cholesterol is carried as part of the plasma lipoproteins, i.e. very-low-density lipoproteins (VLDL), low-density lipoproteins (LDL) and high-density lipoproteins (HDL), more attention has been given in recent studies to the levels of cholesterol as part of the individual lipoproteins. According to our present information, the study of the lipoprotein fractions is more effective for our understanding of atherosclerosis than the study of serum cholesterol and serum triglycerides alone. Among the cholesterol-carrying lipoproteins, LDL contains most of the circulating cholesterol, and LDL cholesterol is highly correlated with the total cholesterol level (r = 0.84 for men and 0.88 for women [8]). It is therefore obvious that the relationship of coronary heart disease with LDL cholesterol [9] is similar to that with total cholesterol. In several epidemiologic studies, the significance of LDL as a predictor of coronary heart disease is well documented [10].

In light of this observation, one could speculate that the atherogenic mechanism of LDL is simply associated with increased filtration of this lipoprotein from blood plasma into the intima of blood vessels. According to several new developments, however, another theory appears to provide a more adequate explanation. *Brown and Goldstein* [13] have shown that under physiological conditions, LDL is bound to specific surface receptors of peripheral nonhepatic and nonphagocytic cells. After internalization of LDL, the endogenous cholesterol synthesis of the receptor cell is suppressed by cholesterol which is released from LDL by lysosomal cholesterol ester hydrolase. Furthermore, there is a negative feedback between the uptake of LDL cholesterol and the number of cell surface receptors. From these observations it may be concluded that LDL uptake normally is feedback-regulated in peripheral nonhepatic cells. Disturbances in this mechanism, i.e. an insufficient number of cell surface receptors, lead to a reduced uptake of LDL by the specific LDL-receptor-mediated pathway, to high LDL plasma concentrations and to enhanced synthesis of endogenous cholesterol. The elevated plasma LDL level may be compensated for by an enhanced uptake of possibly altered LDL by phagocytic cells [14].

It is well known that persons homozygous for familial hypercholesterolemia have an increased risk of coronary heart disease with clinical manifestations at ages younger than 20 [11]. A possible explanation for this observation appears to be the complete lack of functioning LDL cell surface receptors of peripheral cells and a secondary elevation of serum LDL. In persons heterozygous for this disease, only 50% of specific LDL cell surface receptors may be found [12]. The strong association between LDL metabolism and coronary heart disease in familial hypercholesterolemia suggests that receptor-mediated control mechanisms are important in cellular cholesterol balance. The excess of LDL in the plasma of these patients might be scavenged by macrophages and, as part of the arterial cells, they may first undergo cholesterol ester accumulation and foamy degeneration.

Role of HDL in Atherogenesis

About 25% of the total serum cholesterol is carried in the HDL fraction. In contrast to the atherogenic role of LDL, HDL appears to protect against coronary heart disease; in several epidemiologic studies an inverse proportionality of HDL cholesterol to coronary heart disease has been demonstrated [15–19]. Furthermore, it was found that premenopausal women who have higher levels of HDL cholesterol than men are relatively immune to coronary heart disease [20]. *Glueck et al.* [21] have shown that in families in which more than one generation lived to age 80 or 90, and in which morbidity from myocardial infarction is decreased, high levels of HDL can be observed. In light of these observations, it has been concluded that HDL cholesterol appears to be a powerful indicator in prediction of risk. The possible antiatherogenic mechanism of HDL may be associated with the metabolic and physiological function of this lipoprotein. According to *in vitro* studies with skin fibroblasts or arterial smooth muscle cells [22], HDL may promote the transport of cholesterol out of these cells. *Carew et al.* [23] and *Stein et al.* [24] have shown that HDL may compete with LDL for binding and uptake at the cellular receptor site. Plasma HDL, however, is not a uniform compound, but contains several subclasses with different composition and turnover rates (HDL_1, HDL_2, HDL_3). It is particularly noteworthy that the majority of apoprotein E found in HDL is carried in the HDL_1 subclass, and that this apoprotein and apoprotein B bind to the same high-affinity cell surface receptors on human fibroblasts [25]. Thus, tissue culture experiments evaluating the possible role of HDL in cholesterol efflux cannot be interpreted with certainty in the absence of knowledge on the quantitative contribution of apoprotein E to the regulation of cellular cholesterol metabolism. Similarly, the effect of diet and drugs on HDL concentration and metabolism should include a consideration of apoprotein E as a constituent of HDL_1.

A further critical observation with respect to the relationship of HDL and coronary heart disease is that HDL cholesterol is negatively correlated with the level of serum triglycerides [16]. Epidemiological studies have suggested that triglycerides themselves may not be atherogenic and that VLDL cannot be used as an independent indicator of risk [8]. On the other hand, it is the experience of physicians that many individuals with high triglycerides may have an increased coronary risk, for instance patients from families with familial combined hypertriglyceridemia [26], and it has not been shown that those individuals necessarily have low HDL cholesterol concentrations. In view of the strong metabolic relationship between VLDL and HDL metabolism [27], the contribution of VLDL triglyceride and HDL cholesterol to the coronary risk may differ among individuals and should be carefully investigated in the future.

Based upon the hypothesis of HDL as a critical molecule in cholesterol removal from cells, one should expect that patients with Tangier disease (absence of HDL from plasma) suffer from cholesterol accumulation in the blood vessel wall and early atherosclerosis. However, accelerated atherosclerosis appears to be unusual in families with Tangier disease; in fact, clinical and morphological observations in several adult patients [28] exclude an increased risk of coronary heart or other vessel disease. A possible explanation for this puzzle can be drawn from the recent observation that these patients carry certain lipoproteins in their postprandial plasma which have the ability to specifically interact with cellular cholesterol. It could be demonstrated that Tangier chylomicrons undergo regular lipolysis in postprandial plasma and that, in the absence of a preformed HDL pool, chylomicron surface remnants accumulate [29]. Electron microscopically, these remnants of chylomicron catabolism appear as discoidal multilamellar aggregates and account for a significant proportion of the uptake of labeled cholesterol from tissue fibroblasts contained in postprandial Tangier plasma. It is therefore possible, though not proven, that surface remnants of chylomicrons also take up cholesterol from cells *in vivo* (fig. 1).

Tangier disease may further provide a unique model which contributes to our understanding of cholesterol metabolism in peripheral body cells. Cholesteryl ester storage in these patients is confined to tissue macrophages (histiocytes), Schwann cells, nevus cells and intestinal smooth muscle cells; other cells of Tangier tissues do not accumulate excessive amounts of lipid [30, 31]. The morphological findings imply that those cells which do not store lipid at least quantitatively differ from cholesteryl ester storage cells (e.g. histiocytes) with respect to cholesterol uptake and removal mechanisms (fig. 1). It is suggested that tissue macrophages which do not possess LDL receptors [14] are particularly vulnerable to cholesteryl ester accumulation.

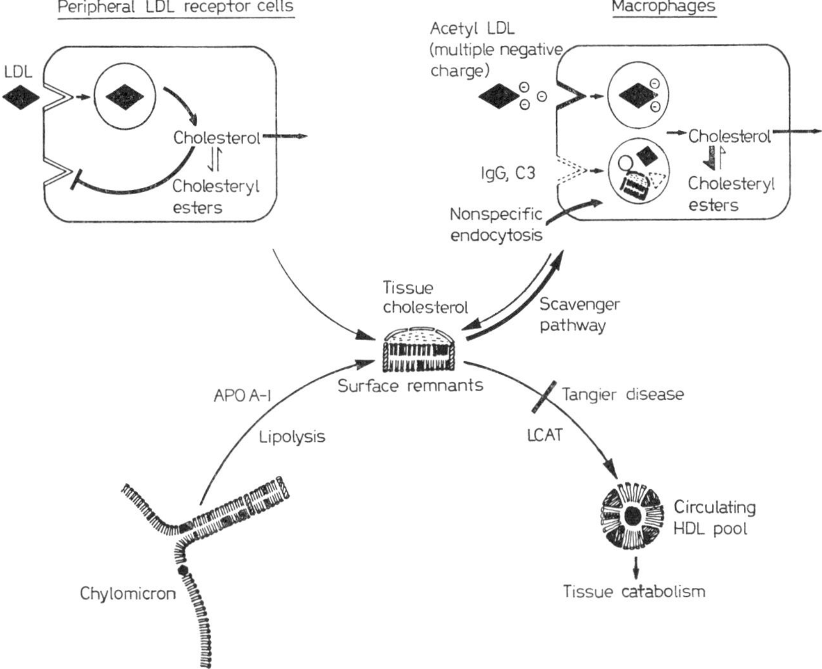

Fig. 1. Tangier disease. Hypothetical relationship between cellular cholesterol metabolism and chylomicron metabolism.

In Tangier disease, the feedback regulatory mechanisms of LDL receptor sites are operative and cholesterol input into the cell is controlled. According to the scheme outlined in figure 1, excess cellular cholesterol in Tangier patients is evacuated through the surface remnants of chylomicrons into macrophages which, as a consequence, accumulate excessive amounts of cholesterol. This mechanism would both explain the absence of early atherosclerosis and the sites of lipid storage in these patients.

The suggested role of chylomicron surface remnants as a mechanism for the protection of Tangier patients from early atherosclerosis may have important general implications. It is well possible that regular lipolysis of triglyceride-rich particles, rather than high serum levels of HDL cholesterol *per se*, is a prerequisite for tissue cholesterol mobilization. The evidence derived from Tangier disease argues against an essential role of mature HDL and implicates HDL precursors in atherogenesis.

References

1   Bronte-Steward, B.; Keys, A., and Brock, J.F.: Serum-cholesterol, diet and coronary heart disease: inter-racial survey in Cape peninsula. Lancet *ii:* 1103–1108 (1955).

2   Doyle, J.T.; Heslin, A.S.; Hillebol, H.E.; Formel, P.F., and Korns, R.F.: Measuring the risk of coronary heart disease in adult population groups. A prospective study of degenerative cardiovascular disease in Albany: report of three years experience. I. Ischemic heart disease. Am. J. publ. Hlth *47:* suppl. April, pp. 25–32 (1957).

3   Chapman, J.M.; Goerke, L.S.; Dixon, W.; Loveland, D.B., and Philips, E.: Measuring the risk of coronary heart disease in adult population groups: the clinical status of a population group in Los Angeles under observation for two to three years. Am. J. publ. Hlth *47:* suppl. part II, pp. 33–42 (1957).

4   Keys, A.; Kimura, N.; Kusunawa, A.; Bronte-Steward, B.; Larsen, N., and Keys, M.H.: Lessons from serum cholesterol studies in Japan, Hawai and Los Angeles. Ann. intern. Med. *48:* 83–94 (1958).

5   Stamler, J.; Lindberg, H.A.; Berkson, D.M.; Shaffer, A.; Miller, W., and Poindexter, A.: Prevalence and incidence of coronary heart disease in strata of the labor force of a Chicago industrial corporation. J. chron. Dis. *11:* 405–420 (1960).

6   Paul, O.; Lepper, M.H.; Phelan, W.H.; Dupertuis, G.W.; McMillan, A.; McKean, H., and Park, H.: A longitudinal study of coronary heart disease. Circulation *28:* 20–31 (1963).

7   Keys, A.; Taylor, H.L.; Blackburn, H.; Brozek, J.; Anderson, J.T., and Simonson, E.: Coronary heart disease among Minnesota business and professional men followed fifteen years. Circulation *28:* 381–395 (1963).

8   Kannel, W.B.; Castelli, W.P., and Gordon, T.: Cholesterol in the prediction of atherosclerotic disease. New perspective based on the Framingham study. Ann. intern. Med. *90:* 85–91 (1979).

9   Gofman, J.W.; Lindgren, F.; Elliot, A.; Mantz, W.; Hewitt, J.; Strisower, B.; Herring, B.; Herring, V., and Lyon, T.P.: The role of lipids and lipoproteins in atherosclerosis. Science *111:* 166–171 (1950).

10 Scanu, A.M.: Plasma lipoproteins and coronary heart disease. Ann. clin. Lab. Sci. *8:* 79–83 (1978).

11 Stone, N.J.; Levy, R.I.; Fredrickson, D.S., and Verter, M.S.: Coronary heart disease in 116 kindred with familial type II hyperlipoproteinemia. Circulation *49:* 476–488 (1974)

12 Brown, M.S.; Dana, S.E., and Goldstein, J.L.: Receptor-dependent hydrolysis oJ cholesterol esters contained in plasma low density lipoprotein. Proc. natn. Acad. Sci. USA *72:* 2925–2929 (1975).

13 Goldstein, J.L. and Brown, M.S.: The low density pathway and its relation to atherosclerosis. A. Rev. Biochem. *46:* 897–930 (1977).

14 Goldstein, J.L.; Ho, Y.K.; Basu, S.K., and Brown, M.S.: Binding site on macrophages that mediates uptake and degradation of acetylated low density lipoprotein, producing massive cholesterol deposition. Proc. natn. Acad. Sci. USA *76:* 333–337 (1979).

15 Miller, G.J. and Miller, N.E.: Plasma high density lipoprotein concentration and development of ischaemic heart disease. Lancet *i:* 16–19 (1975).

16 Castelli, W.P.; Doyle, J.T.; Gordon, T.; Hames, C.G.; Hjortland, M.C.; Hully, S.B.; Kagan, A., and Zukel, W.J.: HDL-cholesterol and other lipids in coronary heart disease. The cooperative lipoprotein phenotyping study. Circulation *55:* 767–772 (1977).

17 Berg, K.; Børresen, A., and Dahlén, G.: Serum high density lipoprotein and atherosclerotic heart disease. Lancet *i:* 499–502 (1976).

18 Gordon, T.; Castelli, W.P.; Hjortland, M.C.; Kannel, W.B., and Dawber, T.R.: Diabetes, blood lipids and the role of obesity in coronary heart disease risk for women. Ann. intern. Med. *87:* 393–397 (1977).

19 Rhoads, G.G.; Gulbrandsen, C.L., and Kagan, A.: Serum lipoproteins and coronary heart disease in the population study of Hawaii Japanese men. New Engl. J. Med. *294:* 293–298 (1976).

20 Mjøs, O.D.: High density lipoproteins and coronary heart disease. Scand. J. clin. Invest. *37:* 191–193 (1977).

21 Glueck, C.J.; Fallat, R.W.; Millet, F.; Gartside, P.; Elston, R.C., and Go, R.C.P.: Familial hyperalpha-lipoproteinemia studies in eighteen kindreds, Metabolism *24:* 1243–1265 (1975).

22 Stein, Y.; Glangeaud, M.C.; Fainaru, M., and Stein, O.: The removal of cholesterol from aortic smooth muscle cells in culture and Landschütz ascites cells by fractions of human high density apolipoproteins. Biochim. biophys. Acta *380:* 106–118 (1975).

23 Carew, T.E.; Koschinsky, T.; Hayes, S., and Steinberg, D.: A mechanism by which high density lipoproteins may slow the atherogenic process. Lancet *i:* 1315–1317 (1976).

24 Stein, O.; Weinstein, D.B.; Stein, Y., and Steinberg, D.: Binding, internalization and degradation of low density lipoprotein by normal human fibroblasts from a case of homozygous familial hypercholesterolemia. Proc. natn. Acad. Sci. USA *73:* 14–18 (1976).

25 Bersot, T.P.; Mahley, R.W.; Brown, M.S., and Goldstein, J.L.: Interaction of swine lipoproteins with the low density lipoprotein receptor in human fibroblasts. J. biol. Chem. *251:* 2395–2398 (1976).

26 Brunzell, J.D.; Schrott, H.G.; Motulsky, A.G., and Bierman, E.L.: Myocardial infarction in the familial forms of hypertriglyceridemia. Metabolism *25:* 313–320 (1976).

27 Nikkilä, E.A.: Metabolic regulation of plasma high density lipoprotein concentration. Eur. J. clin. Invest. *8:* 111–113 (1978).

28 Assmann, G.: Tangier disease and the possible role of HDL in atherogenesis; in Paoletti and Gotto, Atherosclerosis reviews, vol. 6, pp. 1–28 (Academic Press, New York, 1979).

29 Assmann, G. and Schmitz, G.: Tangier disease: a biochemical explanation for the absence of early atherosclerosis. Eur. J. clin. Invest. (in press).

30 Ferrans, V.J. and Fredrickson, D.S.: The pathology of Tangier disease. A light and electron microscopic study. Am. J. Path. *78:* 101–158 (1975).

31 Assmann, G. and Schaefer, H.E.: High density lipoprotein deficiency and lipid deposition in Tangier disease; in Carlson, Int. Conf. on Atherosclerosis, pp. 97–101 (Raven Press, New York, 1978).

G. Assmann, Zentrallaboratorium der Medizinischen Einrichtungen der Universität Münster, D-4400 Münster (FRG)

Nutr. Metab. *24* (Suppl. 1): 26–33 (1980)

Effects of HDL Subclasses on 3-Hydroxy-3-Methyl-glutaryl Coenzyme A Reductase in Human Fibroblasts

W.H. Daerr, Sandra H. Gianturco, J.R. Patsch, L.C. Smith and A.M. Gotto

Klinisches Institut für Herzinfarktforschung,
Medizinische Universitätsklinik, Heidelberg

Key Words. Zonal centrifugation · HDL_2 · HDL_3 · Cholesterol biosynthesis · Tissue culture

Abstract. Monolayer cultures of normal human fibroblasts were used to study the effects of the main subclasses of high-density lipoproteins, HDL_2 and HDL_3, on 3-hydroxy-3-methylglutaryl CoA reductase (EC 1.1.1.34) activity. In this system, HDL_3 (d = 1.125–1.210 g/cm³) specifically induced HMG-CoA reductase activity. Evaluation of culture dynamics revealed that enzyme induction was restricted to the stationary phase of growth. When the cells were incubated with HDL_2 (d = 1.063–1.125 g/cm³), suppression of reductase activity was observed. Mixtures of HDL_2 and HDL_3 suppressed reductase activity when HDL_2 was greater than 35% of the total HDL. The suppressive effects of HDL_2 were abolished by treatment with cyclohexanedione and restored by regeneration of the arginyl residues, suggesting an apoprotein-mediated suppressive mechanism. These observations show that the cellular effects of HDL depend upon the stage of cell growth and the ratio of HDL subclasses in HDL as usually isolated.

Plasma high-density lipoprotein (HDL) concentrations appear to be inversely associated with the development of coronary heart disease [CHD; 1–3]. There is strong epidemiologic evidence that the lower their levels the greater the risk of CHD. Regarding the molecular basis for this negative correlation, the beneficial effect of HDL in the pathogenesis of atherosclerosis may be the result of several mechanisms. HDL may modulate the uptake of cholesterol-rich low-density lipoprotein (LDL) by peripheral tissues [4,5] and function as a vehicle for the transport of cholesterol to the liver [6,7]. HDL may also play a role in the clearance of cholesterol from the arterial wall, since HDL in the nutrient medium promote the release of cholesterol from various cells in tissue culture [8–10]. In agreement with the latter concept, HDL have been reported to induce changes in primary cultures of rat hepatocytes and mouse fibroblasts that might be considered compensatory mechanisms for enhanced cholesterol depletion [10,11]. There is, however, little evidence for similar counter-regulatory alterations in normal human fibroblasts. Although HDL have been recently reported to increase the number of cell surface receptors for LDL in human fibroblasts [12], other studies to date have failed to demonstrate a selective effect of HDL on cholesterol metabolism in these cells [13–15].

In evaluating factors that might limit the effect of HDL on sterol synthesis in normal human fibroblasts, one should remember that a great many of the foregoing observations were made on cells that had been grown for some time in the lipoprotein-deficient fraction of serum (LPDS), which itself promotes activation of sterol synthesis. Therefore, the possibility has to be considered that effect of HDL on sterol synthesis may be obscured by the superimposition of the stimulatory effect of LPDS. Another possibility is that HDL subclasses have independent and opposing effects on sterol flux in cultured human fibroblasts. The balance between the actions of HDL subclasses may then determine the net loss of cholesterol from the cells and hence the levels of 3-hydroxy-3-methylglutaryl coenzyme A reductase (HMG-CoA reductase) activity.

The present report summarizes experiments that we carried out to determine how inherent properties of the fibroblast model system and the subclass distribution of HDL influence the effect of HDL on sterol synthesis. In these studies, use was made of the rate zonal ultra-centrifugation technique, which permits the large-scale preparation of the major subclasses of HDL, HDL$_2$ and HDL$_3$.

We began these studies by measuring the activity of HMG-CoA reductase at different stages of the growth cycle in cells that had been preincubated for 42 h with the LPDS. In the absence of HDL, enzyme activity declined by more than 90% from 190 pmol/min/mg during the log phase to 21 pmol/min/mg during the stationary phase. When HDL$_3$ was added to the medium 18 h prior to the determination of enzyme activity the response of logarithmically dividing cells and cells in the stationary phase was quantitatively different: HDL$_3$ did not affect HMG-CoA reductase activity in proliferating cells, but induced enzyme activity 1.5- to 3-fold when cell replication nearly stopped.

The question that is posed immediately is, are these changes in enzyme activity truly specific or do they rather reflect HDL$_3$-mediated reinitiation of cell growth. It should be stressed here that incubation with HDL$_3$ had no effect on DNA and total protein content per dish in confluent cultures

Table I. Response of normal human fibroblasts to HDL_3.

Modification of the media		HMG-CoA Reductase pmol/min/mg	DNA, /ug	Total Protein mg	Acid phosphatase nmol/min/mg	Alkaline phosphatase nmol/min/mg	Viability (trypan blue exclucion) %
Control		26.7	11.8	0.173	17.1	0.22	98
0.1 mM cyclo-heximide		1.6	–	0.162	17.3	–	96
0.1 mM actinomycin		2.0	–	0.133	14.7	0.24	98
HDL_3,	0.320 mg/ml protein	63.4	11.5	0.168	16.3	0.23	96
HDL_3,	0.320 mg/ml + 0.1 mM cyclo-heximide	2.0	–	0.167	15.9	–	97
HDL_3,	0.320 mg/ml + 0.1 mM actino-mycin D	3.5	–	0.111	14.1	0.25	98

Normal human fibroblasts were initially plated at a concentration of 4×10^4 cells per dish in 5 ml of growth medium containing 10% fetal calf serum. After 5 days the monolayers were washed and placed on 2 ml of medium containing 5% LPDS for 24 h. The growth medium was then modified as indicated. At 18 h after modification of the medium, duplicate dishes of each set of experiments were randomly chosen for measurement of enzyme activities, DNA and total protein content per plate and for the determination of cell viability, All values are the means of duplicate determinations from two different dishes. Standard deviations were less than 12% of the mean.

(table I). Likewise, the increase in HMG-CoA reductase activity was associated with enhanced incorporation of acetate into nonsaponifiable lipids in the intact cells, but did not affect incorporation of acetate into saponifiable lipids (table II). These results lend strong support to the concept that reductase activity was stimulated in a specific fashion.

Further experiments have indicated that HDL_3-mediated stimulation of reductase activity was dependent upon HDL_3 concentrations, could be prevented by incubation with the metabolic inhibitors cycloheximide and actinomycin D and was also brought about with dispersions of HDL_3 lipid, but not when the cells were either grown with human serum albumin or apo-HDL apolipoprotein. We feel that the most probable explanation of these

Table II. Effects of HDL$_3$ and LDL on HMG-CoA reductase activity and the synthesis on sterols and fatty acids from 1-^{14}C-acetate in normal human fibroblasts.

Modification of the media	HMG-CoA reductase, mean ± SD pmol/min/mg	cholesterol synthesis, mean ± SD dpm × 10^3/mg/h	Fatty acid synthesis, mean ± SD dpm × 10^3/mg/h
Control	17.0 ± 3.0	2.48 ± 0.21	3.63 ± 0.67
HDL$_3$, 0.320 mg protein/ml	40.0 ± 2.0	4.27 ± 0.32	3.56 ± 0.31
LDL, 0.010 mg protein/ml	1.9 ± 0.2	0.52 ± 0.06	2.90 ± 0.08

Normal human fibroblasts were grown in complete medium containing 10% fetal calf serum. Medium was changed daily. On day 8 the monolayers were washed and placed on fresh medium containing 5% LPDS for 24 h. Indicated quantities of HDL$_3$ and LDL were then added. After further incubation for 18 h, 12.9 /uCi of 1-^{14}C-acetate at 0.1 mM final concentration were added and the incubation continued for 20 min. The cells were then harvested for measurement of HMG-CoA reductase activity and incorporation of radioactivity into cell sterols and fatty acids. All values are means of measurements from triplicate dishes.

results is that the lipoprotein lipid moiety was indispensable for the stimulatory effect of HDL$_3$ and that stimulation of reductase activity required *de novo* enzyme synthesis.

It is striking that the stimulatory effect of HDL$_3$ on HMG-CoA reductase was restricted to the stationary phase of growth. This phenomenon may be related to quantitative variations in the cells' requirement for cholesterol as cell density changes. In tissue culture cells, which are grown for some time in LPDS, the level of reductase activity is defined to a great extent (a) by the degree of cholesterol efflux into the surrounding medium [16, 17] and (b) by the rate of new plasma membrane synthesis [18]. Since the number of dividing cells and hence the rate of membrane synthesis is large in the log phase of growth, induction of HMG-CoA reductase at this stage is presumably maximal. In confluent resting cultures, however, loss of cellular cholesterol – rather than cell division – may be the prevailing effector for reductase stimulation. At this stage the degree of stimulation may depend primarily upon the effectiveness of the acceptor system in the medium that can stimulate the release of cholesterol from the cell. That a direct relationship exists between sterol loss and phospholipid content of the acceptor system in the medium has been recently demonstrated [16, 17]. Therefore we suggest that it is the high content of phospholipid of HDL_3 which at confluency allows the substantially greater induction of HMG-CoA reductase than seen with LPDS.

In contrast to the increase in reductase activity with HDL_3, a concentration-dependent suppression of reductase was observed when cells were incubated with HDL_2. Interestingly, maximal inhibition which could be achieved varied between experiments. In numerous experiments with two different preparations of HDL_2 maximal inhibition ranged from 17 to 52%. The inhibitory effect was less marked when enzyme activity was low in control cells, indicating that the age of the monolayer culture may have played a role in the extent of enzyme inhibition.

Wath accounts for the inhibitory ability of HDL_2 is not clear at the present time. Although it is possible that enzyme suppression could have been caused by contamination with LDL, this seems unlikely since HDL_2 did not contain immunochemically detectable apolipoprotein B. Also, when individual fractions of the HDL_2 peak from the zonal rotor effluent were tested for their ability to suppress HMG-CoA reductase activity, they were found to be equally inhibitory. This observation is consistent with the coincidence of the inhibitory activity with HDL_2. Another possibility is that the inhibitory effect of HDL_2 was associated with the presence of a minor, but potent, subfraction of lipoproteins which contains the apolipoprotein E and which is isolated with the high-density fraction d 1.063 to 1.125 [19]. This HDL subfraction resembles HDL_c with respect to its capacity to bind to the specific LDL cell surface receptors of normal human fibroblasts [19] and hence might be expected to suppress HMG-CoA reductase activity. A number of experiments suggested that apolipoprotein E may indeed be involved in the HDL_2-mediated inhibition of reductase activity. First of all, HDL_2 failed to suppress HMG-CoA reductase in receptor-negative fibroblasts. In the second place, rather than suppression, the HDL_2 lipid extract elicited induction of enzyme activity, thereby clearly demonstrating the importance of the apolipoprotein moiety for the inhibitory effect of HDL_2. Finally, the inhibitory ability of HDL_2 was reduced by more than 80% following treatment with cyclohexanedione, a reagent known to abolish the biological activity of apolipoprotein-E-containing lipoproteins. Whether HDL_2 also had an inhibitory effect by virtue of its high cholesterol content relative to that of HDL_3 is unclear from our data. In any event, the inhibitory effect of HDL_2 cannot be explained by a nonrandom cellular uptake mechanism alone. A characteristic feature of the lipoprotein-mediated inhibition of HMG-CoA reductase in normal human fibroblasts is that internalization of the lipoprotein particles at specialized sites of binding has to be associated with delivery to the cell of sufficiently large amounts of cholesterol. This requirement may be met by HDL$_2$ but not equally well by HDL$_3$.

So far, these experiments have given us some insight into the variations in sterol synthesis that occur following incubation of normal human fibroblasts with single subclasses of HDL. The last experimental data to be discussed are the effect of mixtures of HDL$_2$ and HDL$_3$ on HMG-CoA reductase activity. These experiments are of particular interest since the subclass composition of HDL is known to be variable [20–23]. Hence, depending upon the inhibitory potency of HDL$_2$, the effect of the total HDL might vary as a function of the ratio of HDL$_2$ and HDL$_3$. To explore this possibility monolayer cultures were incubated with a constant amount of HDL$_3$ and increasing amounts of HDL$_2$. Under these conditions a progressive reduction of enzyme activity and eventually suppression below the level in control cells was indeed observed when the concentration of HDL$_2$ was greater than 35% of the total HDL.

Summarizing, we conclude that not only variations in culture conditions but also differences in the subclass distribution of HDL ought to be considered in the interpretation of studies investigating cellular responses to HDL. The influence of the stage of growth on the effect of HDL$_3$ on HMG-CoA reductase activity and the ability of HDL$_2$ to antagonize the stimulatory effect of HDL$_3$ provide possible explanations for the failure of previous investigations to demonstrate a selective effect of HDL on sterol synthesis in normal human fibroblasts [14, 24]. Further experiments will be needed to delineate the biological significance of the opposite effects of HDL subclasses on HMG-CoA reductase.

References

1   Miller, G.E. and Miller, J.E.: Plasma high density lipoprotein concentration and development of ischaemic heart disease. Lancet *i:* 16–19 (1975).

2   Rhoads, G.G.; Gulbrandsen, C.L., and Kagan, A.: Serum lipoproteins and coronary heart disease in a population study of Hawaii Japanese men. New Engl. J. Med. *294:* 293–295 (1976).

3   Gordon, T.; Castelli, W.P.; Hjortland, M.C.; Kannel, W.B., and Dawber, T.R.: High density lipoprotein as a protective factor against coronary heart disease. Am. J. Med. *62:* 707-714 (1977).

4   Stein, O. and Stein, Y.: Surface binding and interiorization of homologous and heterologous serum lipoproteins by rat aortic smooth muscle cells in culture. Biochim. biophys. Acta *398:* 377–384 (1975).

5 Miller, N.E.; Weinstein, D.B.; Carew, T.E.; Koschinsky, T., and Steinberg, D.: Interaction between high density and low density lipoproteins during uptake and degradation by cultured human fibroblasts. J. clin. Invest. *60:* 78–88 (1977).

6 Glomset, J.A.: The plasma lecithin: cholesterol acyltransferase reaction. J. Lipid Res. *9:* 155–167 (1968).

7 Halloran, L.C.; Schwarz, C.C.; Vlahcevic, Z.R.; Nisman, R.M.; Swell, L., and Goren, R.: Evidence for high-density lipoprotein-free cholesterol as the primary precursor for bile-acid synthesis in man. Surgery, St Louis *84:* 1–7 (1978).

8 Bates, S.R. and Rothblat, G.H.: Regulation of cellular sterol flux and systhesis by human serum lipoprotein. Biochim.biophys. Acta *360:* 38–55 (1974).

9 Stein, O.; Vanderhoek, J., and Stein, Y.: Cholesterol content and sterol synthesis in human skin fibroblasts and rat aortic smooth muscle cells exposed to lipoprotein-depleted serum and high density apolipoprotein/phospholipid mixtures. Biochim. biophys. Acta *431:* 347–358 (1976).

10 Stein, O. and Stein, Y.: Removal of cholesterol from fibroblasts and smooth muscle cells in culture in the presence and absence of cholesterol esterification in the medium. Biochim. biophys. Acta *529:* 309–318 (1978).

11 Edwards, P.A.: Effects of plasma lipoproteins and lecithin-cholesterol dispersions on the activity of 3-hydroxy-3-methylglutaryl coenzyme A reductase of isolated rat hepatocytes. Biochim. biophys. Acta *409:* 39–50 (1975).

12 Miller, N.E.: Induction of low density lipoprotein receptor synthesis by high density lipoprotein in cultures of human skin fibroblasts. Biochim. Biophys. Acta *529:* 131–137 (1978).

13 Brown, M.S.; Dana, S.E., and Goldstein, J.L.: Regulation of 3-hydroxy-3-methylglutaryl coenzyme A reductase activity in human fibroblasts by lipoproteins. Proc. natn. Acad. Sci. USA *70:* 2162–2166 (1973).

14 Miller, N.E.; Weinstein, D.B., and Steinberg, D.: Binding, internalization and degradation of high density lipoprotein by cultured normal human fibroblasts. J. Lipid Res. *18:* 438–450 (1977).

15 Goldstein, J.L.; Dana, S.E., and Brown, M.S.: Esterification of low density lipoprotein cholesterol in human fibroblasts and its absence in homozygous familial hypercholesterolemia. Proc. natn. Acad. Sci. USA *71:* 4288–4292 (1974).

16 Henriksen, T.; Evensen; S.A.; Torsvik, H., and Carlander, B.: Human endothelial cells in primary culture – effects of normal lipoproteins on the incorporation of acetate into lipids. Biochim. biophys. Acta *489:* 64–71 (1977).

17 Bersot, T.B.; Mahley, R.W.; Brown, M.S., and Goldstein, J.L.: Interaction of swine lipoproteins with the low density lipoprotein receptor in human fibroblasts. J. biol. Chem. *251:* 2395–2398 (1976).

18 Fogelman, A.M.; Seager, J.; Edwards, P.A., and Popjak, G.: Mechanism of induction of 3-hydroxy-3-methylglutaryl coenzyme A reductase in human leukocytes. J. biol. Chem. *252:* 466–651 (1977).

19 Rothblat, G.H.; Arbogast, L.Y., and Ray, E.K.: Stimulation of esterified cholesterol accumulation in tissue culture cells exposed to high density lipoproteins enriched in free cholesterol. J. Lipid Res. *19:* 350–358 (1978).

20 Patsch, J.R.; Yeshurun, D.; Jackson, R.L., and Gotto, A.M.: Effects of clofibrate, nicotinic acid, and diets on the plasma lipoproteins of a subjet with type III hyperlipoproteinemia. Am. J. Med. *63:* 1001–1009 (1977).

21 Nichols, A.V.: Human serum lipoproteins and their interrelationship. Adv. biol.
    med. Phys. *11:* 109–158 (1967).
22 Anderson, D.W.; Nichols, A.V.; Pan, S.S., and Lindgren, F.T.: High density lipo-
    protein distribution. Resolution and determination of three major components in a
    normal population sample. Atherosclerosis *29:* 161–179 (1978).
23 Patsch, J.R.; Gotto, A.M., Olivecrona, T., and Eisenberg, S.: Formation of high
    density lipoprotein$_2$-like particles during lipolysis of very low density lipoproteins *in
    vitro.* Proc. natn. Acad. Sci. USA *75:* 4519–4523 (1978).
24 Weinstein, D.B.; Carew, T.E., and Steinberg, D.: Uptake and degradation of low
    density lipoprotein by swine arterial smooth muscle cells with inhibition of choleste-
    rol synthesis. Biochim. biophys. Acta *424:* 404–421 (1976).

Dr. W. Daerr, Klinisches Institut für Herzinfarktforschung an der Medizinischen
Universitätsklinik, Bergheimerstrasse 58, D-5900 Heidelberg (FRG)

Nutr. Metab. *24* (Suppl. 1): 34–44 (1980)

High and Low High-Density Lipoproteins: Clinical Implications

P.Avogaro, G. Cazzolato, G. Bittolo Bon and G.G. Quinci

Unit for Atherosclerosis, Hyperlipaemias and Diabetes,
Regional General Hospital, Venice

Key Words. Lipoproteins · Apolipoproteins · HDL-cholesterol · Hypercholesterolaemia · Atherosclerosis · Diabetes mellitus · Obesity · Familial hyper-alpha-cholesterolaemia · Ethanol

Abstract. High-density lipoproteins (HDL) are macromolecular complexes of lipids and proteins which float in the ultracentrifuge between densities of 1.063 and 1.21 g/l. A major division of HDL into two subclasses according to their specific density has been proposed: HDL_2 (d = 1.063–1.125) and HDL_3 (d = 1.125–1.21). The division is supported by the different proportions of lipids and apoproteins in the two subclasses and by evidence suggesting that the protective effect of HDL lies mostly in the HDL_2 subclasses. Recent epidemiological, experimental and clinical data have supported a probable role or HDL as an anti-risk factor due to their negative correlation with the prevalence of ischaemic heart disease. Some of these data are presented and discussed in this paper.

High-Density Lipoprotein as an Anti-Risk Factor

Clinical Data

In 1975, *Glueck et al.* [6] and *Avogaro et al.* [7] described several families affected with a previously unknown familial type of hyper-high-density lipoproteinaemia-(α)-cholesterolaemia [hyper-HDL-(α)-cholesterolaemia]. The anomaly appears to follow an autosomal-dominant pattern of inheritance. In the families of *Glueck et al.,* the total plasma cholesterol values were normal; in our series, some affected members showed total cholesterol values higher than normal. In the affected subjects, most of the cholesterol was carried as HDL cholesterol or it was beyond the confidential limits for normal subjects (>70 mg/dl). HDL cholesterol values appeared to be normal [6] or lower than normal [7]. In the family described by *Avogaro et al.*

[7], apoB appeared lower and apoA higher than in controls. All subjects appeared to be free of any symptom or sign attributable to a hyperlipaemic state. Analysis of pedigrees did not disclose any premature cardiovascular episodes or any case of sudden death. Studying longevity factors in octogenarians, *Glueck et al.* [8] have observed that life expectancy in years is significantly greater in subjects showing low low-density lipoprotein (LDL) cholesterol and/or high HDL cholesterol. Longevity seemed to be correlated with rare cardiovascular episodes and a two- to three fold reduction in the atherogenic ratio LDL cholesterol/HDL cholesterol.

Epidemiological Data

In 1953 *Barr* [9] studied a series of survivors of myocardial infarction in comparison with a series of normal subjects and recorded much lower levels of cholesterol as α-lipoproteins in the cases of infarction. It appeared that the distribution of cholesterol between lipoproteins was more constantly disturbed than the concentrations of LDL. *Gofman et al.* [4] stressed that the appearance of new cases of ischaemic heart disease in a sample of normal population was inversely correlated to the levels of HDL_2 and HDL_3. 20 years later, *Miller and Miller* [10], collecting data from some epidemiological works found an inverse relationship between HDL cholesterol values and the prevalence of ischaemic heart disease in various population samples. This finding was supported by data from the Framingham study [11]. In a series of 2,815 men and women, a strong inverse correlation (< 0.001) was noted between HDL cholesterol and each major manifestation of coronary heart disease, both in men and women. The same authors found a weaker direct relationship between coronary heart disease and LDL cholesterol.

Experimental Data

Studies of the surface binding of HDL to cultured porcine smooth-muscle cells have emphasized that the binding ability of HDL is only slightly less than that of LDL at the same lipoprotein concentration [12, 13]. When incubated with LDL, the smooth-muscle cells show a net increment in cholesterol content; however, cells incubated with equal or higher concentrations of HDL under comparable conditions show no cholesterol accumulation [13]. It has recently been shown that the active subfraction of HDL which binds to the high-affinity receptor site occurs in the ultracentrifugal density fraction d = 1.063–1.125 (HDL-I) which accounts for only 15% of the entire fraction with d = 1.063–1.21 [14]. Feeding cholesterol to both dogs and humans induces the formation from HDL-I of an HDL particularly rich in cholesterol (50%) named HDL_c [14]. As HDL-I is converted to HDL_c, besides being enriched in cholesterol, it is also characterized by a higher apoE content. One important difference is that HDL containing apoE has a binding power 100-fold more active than LDL containing apoB at the same protein concentration [14]. When confluent multilayers of rat aortic smooth-muscle cells in culture are exposed to human LDL and chloroquine, a marked increase in cellular cholesterol ester results [15]. Removal of the accumulated cholesterol ester is greatly enhanced by a mixture of high-density apolipoprotein-spingomyelin [15].

High-Density Lipoprotein According to Age and Sex

Barr [9] stressed that the HDL cholesterol in human infants was as high as in animals resistant to atherosclerosis; as α-lipoproteins decline, humans become susceptible. The same author recorded equal concentrations of cholesterol in young men and women; females, however, showed a greater percentage of total cholesterol in α-lipoprotein. The plasma level of HDL cholesterol was determined in a cohort of 1,025 men and 1,445 women [11]. HDL cholesterol is substantially higher in women than in men. There is no discernible trend for men while for women there is a slight decrease from age 50 to age 80. Recently, we [16] have recorded the levels of HDL cholesterol in two series of 100 subjects each, males and females. HDL cholesterol values were higher in females than in males for the age group 20–49 years while values for the age group 50–69 were very close. Data concerning the influence of age and sex on protein content of HDL are scanty. In the studies performed up to now, values of $apoA_1$ appear to be higher in females than in males [16–18], the difference, however, is significant only in two reports [16, 18]. In both sexes, we have observed a trend toward higher values of $apoA_1$ with advancing age [16].

Congenital Hyper-HDL and Hypo-HDL: Clinical Data

Congenital Hyper-HDL
The clinical aspects and the biochemical characteristics have been outlined above.

Congenital Hypo-HDL – Tangier Disease
Tangier disease is a rare hereditary disorder. Clinical features include orange-coloured tonsils, splenomegaly and peripheral neuropathy. In the

foam cells of many tissues there is storage of cholesterol esters. The absence of normal HDL-α-lipoprotein is the biochemical marker of the disease [19, 20]. There are, moreover, fasting chylomicronaemia, hypertriglyceridaemia and normal plasma cholesterol/cholesterol ester ratio. In quantitative terms, apoA$_I$ and apoA$_{II}$ are reduced to less than 1% and to 5–7%, respectively [16]. The normal ratio apoA$_I$/apoA$_{II}$ decreases from 2:1 to 1:10. The composition of very-low-density lipoproteins (VLDL) in Tangier disease is close to normal with the only exception of a reduced apoC level. Total LDL is reduced from about 50% up to one fifth of the normal amount; this lipoprotein class is abnormally high in triglycerides at the expense of cholesterol esters [20]. Despite the low levels of HDL and apoA$_I$, the patients affected with the disease do not suffer from cardiovascular episodes. It is likely that this is partly due to the low levels of the atherogenic LDL and to the sequestration of cholesterol ester in histiocytes.

Conditions Characterized by Hyper-HDL-(α)-Cholesterolaemia

Ethanol is known as a powerful inducer of hyperlipaemia in both animals and humans [21–23]. An increase in electrophoretic α-lipoprotein following acute ethanol intoxication has been recorded by *Johanson and Laurell* [21]. In a study on lipoprotein classes performed with a zonal rotor [23], we could demonstrate that ethanol induces a rather complex effect: 'areas' corresponding to VLDL, intermediate-density lipoproteins (IDL), LDL, HDL$_2$ and HDL$_3$ are increased; cholesterol percentage decreases significantly in all classes: triglyceride concentration increases in all fractions, while protein percentage increases in HDL$_2$ and HDL$_3$ and decreases in VLDL, IDL and LDL. Total apoB decreases while apolipoproteins soluble in TMU (apoC, apoA, apoE) increase. Even chlorinated hydrocarbon *pesticides* can induce an increase in HDL concentration in exposed subjects [24].

Physical activity has been correlated to high levels of HDL cholesterol [25–29]. Even populations characterized by sustained physical activity as Jamaican farmers [26] and Eskimos [29] show substantially higher HDL cholesterol levels. This relationship has been recently questioned by *Lipson et al.* [30]. Relying on the data derived from an exercise program, these authors affirm that when diet and weight are controlled, exercise *per se* does not elevate HDL cholesterol concentration and may actually decrease it, with a fall in total cholesterol. The higher HDL cholesterol levels found in trained athletes may be due, therefore, to other aspects of their metabolism or life style.

Body weight has an inverse relationship to both HDL cholesterol and apoA$_I$ [31]. Following a *weight loss,* increased levels of HDL have been stressed while VLDL and LDL concentrations decrease [32]. In a series of 30 obese subjects we have observed that patients do not differ from controls in the values of VLDL cholesterol, VLDL triglycerides, LDL cholesterol, total apoB, LDL-apoB, VLDL-apoB while they significantly differ in values of apoA$_1$ and HDL cholesterol [33]. The weight loss in the obese following a hypocaloric dietary treatment (1,000 kcal; P 20; F 35; CHO 4; cholesterol < 300 mg for 40–50 days) induced a decrease in cholesterol (total and LDL) and triglycerides (total and VLDL), and a significant increase in apoA$_I$ and HDL cholesterol [33]. ApoA$_I$ and HDL cholesterol following diet, did not reach normal values, however, although VLDL triglycerides were reduced to levels significantly lower than in controls. Other factors, therefore, besides the known inverse relationship between VLDL and triglycerides, are at the origin of the decreased HDL cholesterol and apoA$_I$ levels in the obese.

Oestrogens have proved to be effective in rising HDL cholesterol in survivors of myocardial infarction [9] and in œstrogen-deficient post-menopausal women [34].

Both clofibrate and *nicotinic acid* significantly increase plasma HDL [35, 3]. Nicotinic acid induces a 340% increase in the HDL$_2$/HDL$_3$ ratio with a specific enrichment of apoA$_I$ in HDL$_2$ and of apoA$_{II}$ in HDL$_3$ [3].

Conditions Characterized by Hypo-HDL-(α)-Cholesterolaemia

The *hyperlipaemic states* are characterized by reduced levels of HDL cholesterol [36]; the lowest level has been observed in type I whereas in type II the decrease is much smaller. In homozygotes, type IIA reduced levels of apoA$_I$ have also been recorded [37]. Reduced HDL cholesterol [38, 39] and apoA$_I$ [40–43] levels have been recorded in patients affected with *coronary atherosclerosis*. Studies performed by using zonal rotor ultracentrifugation have shown that in atherosclerotic patients both HDL$_2$ and HDL$_3$ decrease [42]. Studies of HDL by isoelectrofocusing have stressed a decrease in peptides between isoelectric points 5.8 and 5.15 then related to apoE and apoA$_I$ [41] while the peptides corresponding to the C family are unmodified. In these patients, therefore, there is a reduction in the ratio apoA$_I$ + apoE/ apoC$_{III}$ + apoC$_{III}$.

In survivors of myocardial infarction we could observe that $apoA_I$, apoB and the ratio $apoA_1/apoB$ are as good as lipids in discriminating between patients and controls below the age of 50; beyond this age, lipids loose their values while apolipoproteins keep their discriminating power up to the eighth decade [43]. In atherosclerotic patients, apoB appears a better discriminator than $apoA_1$ while the ratio $apoB/apoA_I$ appears as the most useful discriminator.

The same biochemical findings have been recorded in patients affected with angina pectoris [41]; the recorded variations, therefore, are likely to be referred to the atherosclerotic process *per se* and not to the miocardial infarction. It is noteworthy, however, that in man following an acute myocardial infarction there is a decrease in $HDL\text{-}apoA_I$ while HDL cholesterol remains unchanged [44]. Studies are needed to solve the problem whether these variations of HDL in atherosclerotic patients are genetically determined or acquired in the life time due to environmental factors. Reduced HDL cholesterol and $apoA_I$ levels due to *overweight* have been stressed. Studies on patients with *diabetes mellitus* have offered conflicting data. Hyper-α-lipoproteinaemia [45] and high HDL cholesterol levels have been recorded in insulin-independent diabetics [34]. Low HDL cholesterol levels [9, 46] and normal apoA values [46] have also been noted. In juvenile diabetes, normal HDL cholesterol values have been found [47, 48] without a relationship with the degree of diabetic microangiopathy [48]. No correlation has been found between levels of HB A_{1c} or blood glucose and HDL cholesterol levels [47, 49]; low HDL cholesterol levels have been related to the insulin doses [48]; others report a rise in the level of HDL cholesterol with glucose control by insulin [34, 50]. A lowering in HDL cholesterol has been recorded following the administration of sulphonylureas [51] while other researches did not find any change in HDL cholesterol following both insulin and hypoglicaemic agents [51]. HDL cholesterol levels decrease following treatment with *progestative* [52], *androgen* [53] or oral *contraceptives* [54]. Studies on the effects of diets enriched with polyunsaturated fatty acids have given conflicting results. A diet especially enriched with *polyunsaturated* fatty acids (P/S: 4) reduced plasma cholesterol (24%), VLDL cholesterol (25%), LDL cholesterol (20%) and HDL cholesterol (33%); there was a concomitant decrease in plasma HDL_2/HDL_3 ratio with a decrease in $apoA_I$ [21%; 3]. In another study utilizing polyunsaturated fatty acids (P/S: 1.5), HDL cholesterol, HDL_2 and HDL_3 remained unchanged [55]. in the Oslo study, the hyperlipaemic patients receiving an isocaloric diet with a P/S ratio of 1.01 for 4 years showed a significant reduction in total cholesterol and an increase in HDL cholesterol [56]. *Hypertriglyceridaemic* conditions are characterized by low levels of HDL cholesterol [36, 57]; this is probably correlated to a defective catabolism of chylomicrons and/or VLDL.

In *liver diseases,* decreased concentrations of α- and pre-β-electrophoretic lipoproteins have been recorded [58]. The decreased concentration of HDL has been attributed to an apoA deficiency in binding lipids. An abnormal apoA is present which is dissociated in its peptidic constituents apoA$_I$ and apoA$_{II}$. Moreover, apoA is absent in VLDL.

Clinical Implications

The main interest for HDL is focused on their relationship with atherosclerosis and specifically with coronary heart disease. HDL are essential for the regular conversion and interrelation between VLDL and LDL. During active triglyceride hydrolysis excess surface components, phospholipids, apoproteins C and free cholesterol are transferred to HDL whose functions seem to be a 'c-peptide reservoir and lipid acceptor' [20]. Many experimental data support the view that HDL play a protective role towards the parietal wall. The action of HDL appears rather complex: they both antagonize the uptake and internalization of LDL and promote an efflux of cholesterol from the organismic pools and possibly from the vascular system [12, 14, 15].

Some clinical data support the experimental findings. Few papers describe families whose members show high levels of HDL (α)-cholesterol and seem to be protected against cardiovascular episodes [6, 7]. Other studies stress an inverse correlation between HDL cholesterol levels and the prevalence of ischaemic heart disease [9–11]. Further support is given by studies concerning the relationship of HDL with age, body weight, hormonal status [16–18, 34]; all of them supporting a role of HDL as an anti-risk factor. Other clinical findings contradict this role of HDL, *in primis* the Tangier disease, a pathological condition not aggravated by atherosclerosis despite the extreme reduction of the peptide A$_I$ [19, 20]. A widespread collection of cholesterol esters in foam cells of many viscera except the vascular wall is a peculiar finding of the disease stressing the possibility that HDL are not an essential requisite for all cells for clearing cholesterol [20]. Even data recorded in subjects affected with the various hyperlipaemic states do not properly fit the hypothesis of HDL as a beneficial factor. Type I disease has the lowest level of HDL cholesterol among the various hyperlipaemic states but it is not associated with premature atherosclerosis. For both the Tangier disease and type I, the rarity of premature atherosclerosis is also linked to low levels of LDL typical for both disorders. Whatever the explanation of these situations may be, HDL appear to play a smaller role than LDL which is a key factor in the atherogenetic process.

Also the large body of studies dedicated to diabetes mellitus [34, 45–51] do not disclose any correlative evidence for a role of HDL in explaining the high rate of incidence of atherosclerosis in this disease.

Most studies concerning HDL are based on the determination of HDL cholesterol which is a small constituent of HDL while protein represents a large mass. Further studies concerning the protein moeity of HDL or HDL as a whole are needed for a better understanding of the role of HDL.

References

1   Anderson D.W.: H.D.L. cholesterol: the variable components. Lancet *i:* 819–820 (1978).
2   Cheung, M.C. and Albers, J.J.: The measurement of apolipoprotein A-I and A-II levels in men and women by immunoassay. J. clin. Invest. *60:* 43–50 (1977).
3   Shepherd, J.: The influence of polyunsaturated fat diets and nicotinic acid therapy on the metabolism and subfraction distribution of human high density lipoproteins; in Gotto, Miller and Oliver, High density lipoproteins and atherosclerosis, pp. 193–206 (Elsevier, Amsterdam 1978).
4   Gofman, J.W.; Young, W., and Tandy, R.: Ischemic heart disease, atherosclerosis and longevity. Circulation *34:* 679–697 (1966).
5   Anderson, D.W.; Nichols, A.V.; Pan, S.S., and Lindgren, F.T.: High density Lipoprotein distribution. Resolution and determination of three major components in a normal population sample. Atherosclerosis *29:* 161–179 (1978).
6   Glueck, C.J.; Fallat, R.W.; Millet, F., abd Steiner, P.M.: Familial hyperalphalipoproteinemia. Archs. intern. Med. *136:* 1025–1028 (1975).
7   Avogaro, P. and Cazzolato, G.: Familial hyper-HDL-(α)-cholesterolemia. Atherosclerosis *22:* 63–77 (1975).
8   Glucck, C.I.; Fallat, R.W.; Spadafora, M., and Gertside, P.: Longevity syndromes. Abstract. Circulation *51–52:* suppl. II, p. 272 (1975).
9   Barr, D.P.: The George E. Brown Memorial Lecture. Some chemical factors in the pathogenesis of atherosclerosis. Circulation *8:* 641–654 (1953).
10   Miller, C.J. and Miller, N.E.: Plasma high density lipoprotein concentration and development of ischaemic heart disease. Lancet *i:* 16–19 (1975).
11   Gordon, T.; Castelli, P.; Hjortland, M.C.; Kannel, W.B., and Dawber, T.R.: High density lipoprotein as a protective factor against coronary heart disease. Am. J. Med. *62:* 707–714 (1977).
12   Stein, O. and Stein, Y.: Comparative uptake of rat and human serum low density

and high density lipoproteins in rat aortic smooth muscle cells in culture. Circulation Res. *36:* 436–443 (1975).

13 Carew, T.E.; Koschinsky, T.; Hayes, S.B., and Steinberg, D.: A mechanism by which high density lipoproteins may slow the atherogenic process. Lancet *i:* 1315–1317 (1976).

14 Mahley, R.W.; Weisgraber, K.H.; Bersot, T.P., and Innerarity, T.L.: Effects of cholesterol feeding on human and animal high density lipoproteins; in Gotto, Miller and Oliver, High density lipoproteins and atherosclerosis, p. 149 (Elsevier, Amsterdam 1978).

15 Stein, Y.; Glangeaud, M.C.; Fainaru, M., and Stein, O.: The removal of cholesterol from aortic smooth muscle cells in culture and Landschutz ascites cells by fractions of human high-density apolipoproteins. Biochim. biophys. Acta *380:* 106–118 (1975).

16 Avogaro, P.; Cazzolato, G.; Bittolo Bon, G.; Belussi, F., and Quinci, G.B.: Values of apo-A$_I$ and apo-B in humans according to age and sex. Clinica chim. Acta *95:* 311–315 (1979).

17 Alaupovic, P.; Curry, M.D., and McConathy, W.J.: Electroimmunoassay of serum apolipoproteins in normolipidemic subjects and patients with primary hyperlipoproteinemias; in Schettler, Goto, Hata, and Klose, Atherosclerosis IV, pp. 220–222 (Springer, Berlin 1977).

18 Albers, J.J.; Wahl, P.W.; Cabana, G.V.; Hazzard, W.R., and Hoover, J.J.: Quantitation of apolipoprotein A-I of human plasma high density lipoprotein. Metabolism *25:* 633–644 (1976).

19 Fredrickson, D.S.; Gotto, A.M., and Levy, R.I.: In Stanbury and Fredrickson, The metabolic basis of inherited diseases; 3rd ed., pp. 493–530 (McGraw-Hill, New York 1972).

20 Assmann, G.: The metabolic role of high density lipoproteins: perspectives from Tangier disease; in Gotto, Miller and Oliver, High density lipoproteins and atherosclerosis, p. 77 (Elsevier, Amsterdam 1978).

21 Johansson, H.G. and Laurell, C.B.: Disorders of serum α-lipoprotein after alcoholic intoxication. Scand. J. clin. Invest. *23:* 231–233 (1969).

22 Avogaro, P. and Cazzolato, G.: Changes in the composition and physicochemical characteristics of serum lipoproteins during ethanol induced lipaemia in alcoholic subjects. Metabolism *24:* 1234–1242 (1975).

23 Avogaro, P.; Cazzolato, G.; Bittolo Bon, G., and Belussi, F.: Ethanol lipemia. A doubtful risk factor; in Schettler, Goto, Hata, and Klose, Atherosclerosis IV, pp. 658–664 (Springer, Berlin 1977).

24 Carlson, L.A. and Kolmodin-Heidman, B.: Hyper-α-lipoproteinemia in men exposed to chlorinated hydrocarbon pesticides. Acta med. scand. *192:* 29–32 (1972).

25 Castelli, W.P.; Gordon, T.; Hjortland, M.C.; Kagan, A.; Doyle, J.T.; Hames, C.G.; Hylley, S.B., and Zukel, W.J.: Alcohol and blood lipids. Lancet *ii:* 153–155 (1977).

26 Miller, G.J.; Miller, N.E., and Ashcroft, M.T.: Inverse relationship in Jamaica between plasma high density lipoprotein concentration and coronary disease risk as predicted by multiple risk-factor status. Clin. Sci. mol. Med. *51:* 475–481 (1976).

27 Carlson, L.A. and Mossfeldt, F.: Acute effects of prolonged heavy exercise on the concentration of plasma lipids and lipoproteins in man. Acta physiol. scand. *62:* 51–59 (1964).

28 Morris, J.N.; Chave, S.P.W.; Adam, C.; Sirey C., and Epstein, L.: Vigorous exercise in leisure-time and the incidence of coronary heart disease. Lancet *i:* 333–339 (1973).

29 Bang, H.O.; Dyerberg, J., and Nielsen, A.B.: Plasma lipid and lipoprotein patterns in Greenland West-Coast Eskimos. Lancet *i:* 1143–1145 (1971).

30 Lipson, L.C.; Monow, R.O.; Schaefer, E.; Brewer, H.B.; Lindgren, F , and Epstein, S.E.: Effects of exercise on human plasma lipoproteins. Abstract. Am. J. Cardiol. *43:* 409 (1979).

31 Avogaro, P.; Cazzolato, G.; Bittolo Bon, G.; Quinci, G.B., and Chinello, M.: HDL-cholesterol, apolipoproteins A_I and B, age and index body weight. Atherosclerosis *31:* 85–91 (1978).

32 Wilson, D.E. and Lees, R.S.: Metabolic relationships among the plasma lipoproteins. Reciprocal changes in the concentration of very low and low density lipoproteins in man. J. clin. Invest. *51:* 1051–1057 (1972).

33 Avogaro, P.; Cazzolato, G.; Bittolo Bon, G., and Quinci, G.B.: Variations of plasma lipoproteins and apolipoproteins B and A-I in obese subjects fed with hypocaloric diet. Obesity bariat. Med. (in press, 1979).

34 Nikkila, E.A.: Metabolic and endocrine control of plasma high density lipoprotein concentration. Relation to catabolism of Triglyceride-rich lipoproteins; in Gotto, Miller and Oliver, High density lipoproteins and Atherosclerosis, p. 177 (Elsevier, Amsterdam 1978).

35 Blum, C.B.; Levy, R.I.; Eisenberg, S.; Hall, M., III; Goebel, R.H., and Berman, M.: High density lipoprotein metabolism in man. J. clin. Invest. *60:* 795–807 (1977).

36 Schaefer, E.J.; Levy, R.I.; Anderson, D.W.; Danner, R.N.; Brewer, H.B., Jr., and Blackwelder, W.C.: Plasma triglycerides in regulation of HDL-cholesterol levels. Lancet *ii:* 391–392 (1978).

37 Karlin, J.B.; Juhn, D.J.; Starr, J.L., and Rubenstein, A.H.: Measurement of human high density lipoprotein apolipoprotein A-I in serum by radioimmunoassay. J. Lipid Res. *17:* 30–37 (1976).

38 Carlson, L.A. and Ericsson, M.: Studies in male survivors of myocardial infarction. II. Quantitative and qualitative serum lipoprotein analysis. Atherosclerosis *21:* 435–450 (1975).

39 Lewis, B.; Chait, A.; Oakley, C.M.O.; Wootton, I.D.P.; Krikler, D.M.; Onitiri, A.; Sigurdsson, G., and February, A.: Serum lipoprotein abnormalities in patients with ischaemic heart disease: comparison with a control population. Br. med. J. *9:* 489–493 (1974)

40 Berg, K.; Borresen, A.L., and Dahlen, G.: Serum high density lipoprotein and atherosclerotic heart disease. Lancet *i:* 499–501 (1976).

41 Avogaro, P.; Bittolo Bon, G.; Cazzolato, G.; Quinci, G.B., and Belussi, F.:Plasma levels of apolipoprotein A-I and apolipoprotein B in human atherosclerosis. Artery *4:* 385–394 (1978).

42 Avogaro, P.; Cazzolato, G.; Bittolo Bon, G., and Belussi, F.: Levels and chemical composition of HDL_2, HDL_3 and other major lipoprotein classes in survivors of myocardial infarction. Artery (in press).

43 Avogaro, P.; Bittolo Bon, G.; Cazzolato, G., and Quinci, G.B.: Are apolipoproteins better discriminators than lipids for atherosclerosis? Lancet *i:* 901–903 (1979).

44 Avogaro, P.; Bittolo Bon, G.; Cazzolato, G.; Quinci, G.B.; Sanson, A.; Sparla, M.;

Zagatti, G.C., and Caturelli, G.: Variations in apolipoproteins B and A-I in the course of acute myocardial infarction. Eur. J. clin. Invest. *8:* 121–129 (1978).

45 Wille, L.E. and Aarseth, S.: Demonstration of hyper-α-lipoproteinemia in three diabetic patients. Clin. Genet. *4:* 281–285 (1973).

46 Kopes-Virella, M.F.L.; Stone, P.G., and Colwell, J.A.: Serum high density lipoprotein in diabetic patients. Diabetologia *13:* 285–291 (1977).

47 Kennedy, A.L.; Lappin, T.R.J.; Lavery, T.D.; Hadden, D.R.; Weaver, J.A., and Montgomery, O.A.D.: Relation of high density lipoprotein chclesterol concentration to type of diabetes and its control. Br. med. J. *ii:* 1191–1194 (1978).

48 Eckel, R.H.; Albers, J.J.; McLean, E.B.; Wahl, P.W., and Bierman, E.L.: High density lipoprotein cholesterol (HDL-CHOL) and microangiopathy in juvenile onset diabetes mellitus. Diabetes *27:* suppl. 2, p. 447 (1978).

49 Elkels, R.S.; Wu, J., and Hambley, J.: Haemoglobin A_{1c}, blood glucose, and high-density lipoprotein cholesterol in insulin-requiring diabetics. Lancet *ii:* 547–548 (1978).

50   Calvert, G.D.; Mannik, T.; Graham, J.J.; Wise, P.H., and Yeates, R.A.: Effects of therapy on plasma high density lipoprotein cholesterol concentration in diabetes mellitus. Lancet *ii:* 66–68 (1978).

51 Stalenhof, A.F.H.; Demacker, P.N.M.; Lutterman, J.A., and Van't Laar, A.: High density lipoprotein and maturity onset diabetes. Lancet *i:* 325 (1978).

52 Manninen, Y., and Malkonene, M.: Hormones and high density lipoproteins. Lancet *ii:* 1155 (1978).

53 Masarei, J.R.L. and Lynch, W.J.: Lowering of HDL-cholesterol by androgens. Lancet *ii:* 827–828 (1977).

54 Arntzenius, A.C.; Van Gent, A.M.; Van der Woort, H.; Stegerhoek, C.I., and Styblo, K.: Reduced high density lipoprotein in women aged 40–41 using oral contraceptives. Lancet *i:* 1223–1227 (1978).

55 Rao and Bacchus: Quoted by Lewis et al. (39).

56 Hjermann, I.; Enger, S.C.; Helgeland, A.; Holme, I.; Leren, P. and Trygg, K.: The effect of dietary changes in high density lipoprotein cholesterol. The Oslo Study. Am. J. Med. *66:* 105–109 (1979).

57 Furman, R.H.; Sanber, S.R.; Alaupovic, P.; Bradford, R.M., and Howard, R.P.: Studies on the metabolism of radio-iodinated human serum alphalipoprotein in normal and hyperlipidemic subjects. J. Lab. clin. Med. *63:* 193–199 (1964).

58 Seidel, D.; Greten, H.; Grisen, H.P.; Wengeler, H. and Wieland, H.: Further aspects on the characterization of high and very low density lipoproteins in Patients with liver disease. Eur. J. clin. Invest. *2:* 259–364 (1972).

Prof. P. Avogaro, MD, PhD, Ospedale regionale, Divisione medica II, Centro regionale per l'Aterosclerosi, le Iperlipemie e il Diabete, I-30100 Venice (Italy)

Nutr. Metab. *24* (Suppl. 1): 45–49 (1980)

Recent International Changes in Coronary Heart Disease Mortality

Frederick H. Epstein

Institute of Social and Preventive Medicine, University of Zürich, Zürich

Key Words. Coronary heart disease · Atherosclerosis · Mortality, Epidemiology · Prevention

Abstract. Monitoring secular trends in mortality as well as morbidity from coronary heart disease (CHD) is important in order to relate favourable or unfavourable habits of daily living to the risk of disease, with a view toward effective preventive measures. A marked decline in CHD mortality has been observed in the United States during the past decade. Similar, but smaller, reductions have been recorded in a few other countries but the overall international tendency is toward unchanged or increasing mortality. There is a suggestion for countries with initially higher rates to register declines while the reverse tends to occur for low-mortality countries. The favourable trends in the United States can be linked to improvements in the style of life but parallel information in other countries is largely lacking. International, collaborative efforts to monitor CHD disease rates, living habits and risk factor distributions by country, region and social groups is a major need in the search for effective prevention programmes in the community.

In the search for environmental causes of disease, the detection of relatively short-term secular trends in incidence is important since probably no other finding, except for the results of intervention studies, can provide such direct evidence for the operation of non-genetic influences in predisposing to disease or protecting against it. As a matter of fact, a major impetus for the upsurge of epidemiological research in cardiovascular disease since the end of World War II came from the realization that coronary heart disease (CHD) had become so much more frequent in the preceding decades. Secular trends have generally been upward in most industrialized countries well into the sixties [1]. It came as an unexpected and welcome surprise in the later seventies that the trend for CHD had unmistakenly turned downward [2]. While the previously upward trends provided evidence for the detrimental influence of prevailing living habits on disease risk, the reversal suggests equally powerfully that beneficial factors are at work. It is the purpose of this short report to present a survey of recent international trends and compare and contrast them with the recent developments in the United States.

International Mortality Trends

At a recent conference on the decline of CHD mortality in the United States, held at the National Heart, Lung and Blood Institute in Bethesda, Md., USA, a survey of international trends was also presented [3]. In addition to the United States, declining trends, though to a lesser degree, have been observed in Canada and Australia, between 1968 and 1977. Other non-European countries which have registered a decrease in reported mortality are Israel and, surprisingly, Japan. The data from Japan require close scrutiny in terms of total and competing causes of mortality since they are inconsistent with the well-founded belief that increasing 'westernization' is accompanied by an increase in CHD frequency. Among the countries outside Europe with reliable mortality statistics, New Zealand records no change in trend in the last decade. The very high rates among white men in South Africa have decreased between 1970 and 1975 [4].

Inside Europe, secular trends have been variable [3]. In general, tendencies up- or downward are similar in men and women. Therefore, the data for men have been taken as being representative in the descriptions which follow. By and large, trends also tend to be similar for the different age groups so that, for the present purpose of giving an overall picture, no age-specific data are shown, as they are provided elsewhere [3]. A detailed age- and sex-specific analysis of international mortality trends in the last 10 years, using regression slopes and taking into account total mortality is in preparation. The available data are summarized (table I), arranging 21 countries or regions in descending order of CHD mortality, as recorded in 1968; a division has been made arbitrarily into thirds with the highest, intermediate and lowest mortality. It is apparent that there is a tendency toward high mortality to fall and low mortality to rise. An interpretation of this finding would be a matter of speculation at such an early stage in trying to delineate the trends, establishing their reality and attempting to explain the differences and similarities.

Table I. Trend of coronary heart disease mortality (arranged in decreasing order) in European men between 1968 and 1977

Region:	Trend	Overall trend		
		none	up	down
Highest mortality				
Scotland	0			
N. Ireland	+			
Finland	−			
England and Wales	0	2	4	1
Ireland	+ +			
Denmark	+			
Sweden	+			
Intermediate mortality				
Czechoslovakia	0			
Norway	−			
Netherlands	0			
Hungary	+	4	1	2
Austria	0			
FRG	0			
Belgium	−			
Lowest mortality				
Bulgaria	+ + +			
Italy	0			
Switzerland	0			
Poland	+ + +	2	5	−
Rumania	+ +			
Yugoslavia	+ + +			
France	+			

0 = no trend; + = upward trend; − = downward trend.

In evaluating data for whole countries, it is important to keep in mind that any trends may differ from one socio-economic group to another and between regions. Differential trends by social class exist in Britain [5]. In view of the fact that CHD mortality frequently shows intranational variations, it is likely that secular trends may also differ regionally.

It cannot be overemphasized that current information is limited to mortality. There are no reliable data in any country on changes in incidence.

A decline in mortality could conceivably be due to a reduction in the lethality of acute myocardial infarction or in the frequency of recurrence of infarction among survivors of an acute attack. There are international plans to institute monitoring systems to yield this missing information.

Explanations for Secular Trends in Mortality

In the United States, the observed reduction in mortality is likely to be due, at least in part, to a reduction in the frequency of elevated risk-factor levels, even though improvements in the treatment of coronary heart disease may play an additional role [6]. There has been a reduction in the consumption of saturated fats, with some increase in polyunsaturated fat intake, which is probably responsible for the observed decrease in mean serum cholesterol levels. Likewise, there has been a marked improvement in the detection and effective treatment of hypertension. The number of heavy smokers has declined. In some segments of the population, the level of physical activity during leisure has increased. There is no doubt that these favourable changes have taken place in the United States in recent years, indicating a step toward the primary prevention of coronary heart disease.

It is not possible presently to relate declining secular mortality trends in other countries to changes in risk-factor status and living habits, except in Belgium [7] and possibly Britain [8]. Nevertheless, there has been an increase in the consumption of polyunsaturated fat between the mid-fifties and the mid-seventies in 8 countries belonging to the European community; a decline in the consumption of saturated fats has been observed in only 2 of these 8 countries, however [9]. It will be an important task in the immediate future to collect prospective information on the trends of risk factors and associated living habits on an international level and to study their influence on mortality and morbidity trends. Information is needed not only to explain declining trends but also to account for any lack of change or increasing CHD rates.

Acknowledgement

The international mortality data on which this report is based have been analyzed in collaboration with the Cardiovascular Disease Unit, World Health Organization, Geneva.

References

1 Epstein, F.H. and Krueger, D.E.: The changing incidence of coronary heart disease; in Morgan Jones, Modern trends in cardiology, pp. 17–35 (Butterworths, London 1969).
2 Levy, R.I.: Progress toward prevention of cardiovascular disease. A thirty-year retrospective. Mod. Concepts cardiovasc. Dis. *47:* 103–108 (1978).
3 Epstein, F.H. and Pisa, Z.: International comparisons in ischaemic heart disease mortality; in Havlick and Feinleib, Conference on the Decline in Coronary Heart Disease Mortality, Bethesda 1978 (National Heart, Lung and Blood Institute, Bethesda). NIH Publication No. 79–1610, May 1979.
4 Wyndham, C.H.: Ischaemic heart disease mortality rates in white South Africans compared with other populations. S. Afr. med. J. *54:* 595–602 (1978).
5 Marmot, M.G.; Adelstein, A.M.; Robinson, N., and Rose, G.A.: Changing social-class distribution of heart disease. Br. med. J. *ii:* 1109–1112 (1978).
6 Havlick, R.J. and Feinleib, M.: Conference on the Decline in Coronary Heart Disease Mortality, Bethesda 1978 (National Heart, Lung and Blood Institute, Nat. Inst. of Health, Bethesda). NIH Publication No. 79–1610, May 1979.
7 Joossens, J.V.; Vuylsteek, K.; Brems-Heyns, E.; Carlier, J.; Claes, J.H., et al.: The pattern of food and mortality in Belgium. Lancet *i:* 1069–1072 (1977).
8 Florey, C.V. de; Melia, R.J.W., and Darby, S.C.: Changing mortality from ischaemic heart disease in Great Britain 1968–76. Br. med. J. *i:* 635–637 (1978).
9 Commission of the European Communities: Influence on health of different fats in food. Inf. Agric., No. 40, pp. 28–32 (Commission of the European Communities, Brussels 1977).

Prof. F.H. Epstein, MD, Institute of Social and Preventive Medicine,
University of Zürich, Gloriastrasse 32, CH–8006 Zürich (Switzerland)

Nutr. Metab. *24* (Suppl. 1): 50–64 (1980)

Dietary Effects on Blood Pressure

S. Heyden and C. G. Hames

Duke University Medical Center, Durham, N.C., and
Evans County Heart Study, Claxton, Ga.

Key Words. Potassium/sodium · Prostaglandins · Weight loss-sodium restriction ·
Acquired salt taste · Linoleic acid

Abstract. Low dietary potassium intake, particularly among black hypertensive
patients, appears to contribute significantly to this disease and its major sequellae. The
high sodium intake of industrialized societies and the epidemic prevalence of hypertension
must be considered in the light of a relatively greater urinary potassium excretion as
compared to a low-sodium diet with a lower urinary potassium excretion. There is no
reason to assume that weight reduction *per se* is the major contributor to lowering of
elevated blood pressure levels although claims to this effect have been made. In general,
a high-caloric diet is loaded with sodium while a low-energy diet has a drastically reduced
sodium content. The hypothesis that a higher dietary linoleic acid intake via increased
prostaglandin synthesis may lead to natriuresis and a drop in blood pressure levels is an
interesting development which needs more testing.

Introduction

The major contributions of dietary effects on blood pressure will be
discussed in four sections beginning with promising new leads on the influence
of potassium. This is followed by a brief summary on the present knowledge
regarding sodium. The controversial role of weight reduction independent of
sodium restriction deserves close attention. Finally, speculations about the
potential of polyunsaturated fatty acids in blood pressure lowering will be
offered, based on suggestive evidence from a limited number of animal and
human experiments.

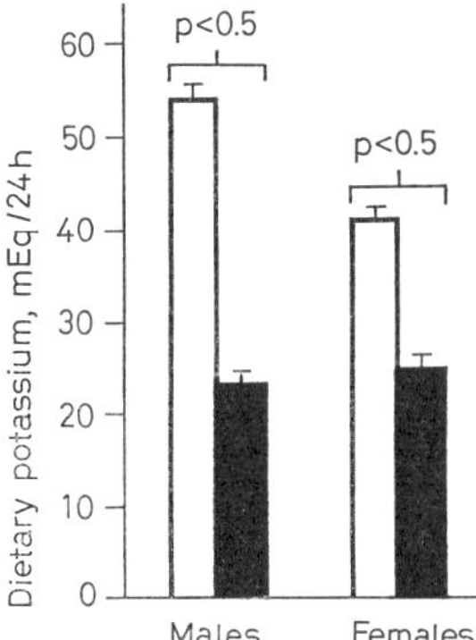

Fig. 1. Different potassium consumption among whites (white bars) and blacks (black bars). Evans County, Ga. [*Grim et al.*, Racial Differences in Blood Pressure in Evans County, Georgia, J. chron. Dis. *33:* 87–94 (1980).

Potassium and Blood Pressure

The ratio of sodium to potassium in the daily food intake may have important implications on *the development of hypertension,* particularly when the salt consumption is high on account of traditional high NaCl usage as in Evans County, Ga. Tobacco chewing (tobacco is a high-sodium product), use of table salt in beer and on any fruit, customary high consumption of ham, canned meats and vegetables add to the excess salt intake among both blacks and whites. But a dietary analysis carried out in 1961 and repeated in 1968 on random samples of the black and white populations revealed a much lower (less than half) potassium intake in blacks, particularly males, compared to whites (fig. 1). There is compelling evidence that for the *entire* population of Evans County the intake of sodium is highly correlated with the prevalence and incidence of hypertension although there is *no real difference in sodium intake* between blacks and whites. The only way to explain the higher prevalence of hypertension among blacks appears to be their much lower potassium intake, caused by low consumption of fresh fruit, salads, vegetables and the traditional long cooking of meats and vegetables with loss of the intracellular potassium into the cooking water [*Grim et al.,* 1980].

The relationship of dietary potassium intake and blood pressure has not been extensively studied in man although observations made in the late 20s have suggested that high potassium intake may lower blood pressure [1].

Table I. Changes in sodium and potassium content of peas

Food (100 g edible portion)	Na, mg	K, mg
Fresh peas	0.9	380
Frozen peas	100	160
Canned peas, liquid poured off	230	180
Add salt, serve with salted butter	(?)	(?)

[From *Meneely and Battarbee,* 15]. Goal 6: Reduce salt consumption by about 50–85% to about 3 g/day.

Studies conducted in the United States [11, 12] have suggested that blood pressure tended to be related to the urinary sodium-potassium ratio. Animal studies [13, 14] have shown that the hypertensinogenic effects of sodium chloride may be lessened by feeding increased amounts of potassium chloride in the diet. Similar results have been reported in spontaneously hypertensive rats and *Grollman and Grollman* [6] were able to induce *de novo* hypertension in the offspring of rats fed a low potassium diet during pregnancy. The relationship between a low potassium diet during pregnancy and the blood pressure of offspring in man has not been investigated, but this conceivably could be operative in the black population *Grim et al.* [5a].

Whereas the human dietary sodium intake may be 3–6 times higher than the potassium intake, quite the reverse is the case with the natural diet of lower animals. The carnivores consume 4–5 times as much potassium as sodium and the herbivores consume even 12–20 times as much potassium as sodium. The average human requirement for sodium is only about 0.25 g [15]. Thus the desire for salt is neither important nor a physiological necessity but an acquired taste. *In contrast, the requirement for potassium is 2.5 g a day.* Common food-processing practices bring about marked changes in sodium and potassium contents. Canning and freezing of peas, for example, deplete them of potassium and increase their sodium contents. Canned peas thus contain 225 times as much sodium as the fresh product and more than half of the potassium is gone (table I). Sodium intake is more and more determined by the food processors rather than by the individual.

There is an extensive literature which documents that individuals who live in primitive communities with a very low sodium intake and a high potassium intake neither develop hypertension nor experience a rise in blood pressure with age. These data show that larger quantities of sodium than those ingested by these primitive tribes are necessary for hypertension to occur; this is reasonably clear, particularly in migrants who have moved from rural to urban dwellings. Very few studies have been done using potassium in the treatment of hypertension. Potassium is a natriuretic agent, and the above-mentioned primitive tribes ingest not only a low-sodium but a very high-potassium diet. *Kempner's* [10] rice diet is a very low-sodium (and rather high-potassium) regimen, and is relatively successful considering the severe cases of hypertension who are being treated. Studies in progress suggest – but not yet unequivocally – hypotensive effects of potassium loading in hypertensive patients who are not on diuretic medication and, therefore, not in hypochloremic alkalosis.

Sodium and Blood Pressure

Ram and Kaplan [17] described experiments involving different drug regimens for hypertension with either normal sodium intake (205 mEq/day) or sodium restriction (64 mEq/day). 'Blood pressures measured hourly during waking hours were lower with the low sodium diet . . . Modest sodium restriction reduces potassium loss and improves antihypertensive potency of diuretics. . . Plasma renin activity and urinary aldosterone rose about 3-fold with diuretic and low sodium diet and 2-fold with diuretic and normal sodium diet.'

Langford [personal commun., 1979], showed that 50 mg of hydrochlorothiazide, given to hypertensive patients on a 50-mEq sodium diet, was as effective in lowering blood pressure as 100 mg hydrochlorothiazide given to the same individuals taking a 100-mEq sodium diet. The important difference between the two treatments was that the higher dose of sodium and hydrochlorothiazide was associated with a lower serum potassium level and a higher urinary potassium excretion. *Langford* stated that no matter what level was picked as 'goal' blood pressure, unsuccessful participants in a hypertension intervention study excreted more sodium than successful ones.

A large sodium intake can thus either negate the effects of the antihypertensive therapy or make a stronger one necessary. Some of the side-effects of the drugs obviously will be greater. *Langford* demonstrated that the sodium depletion produced by the diuretic therapy may increase salt hunger, therefore inspiring the patient to eat more salt, which will partially negate the effects of the therapy.

In February 1977, the McGovern Report published as *Dietary Goal No. 6* salt reduction of the American diet to an intake of 3 g/day NaCl.

On April 18, 1977, a nutritionist, Mr.*White* published a position paper of the American Medical Association (AMA). Unfortunately, his biased views were not reviewed by an expert panel of nutritionists and physicians and no meeting of the AMA Council on Foods and Nutrition was called together. Instead, the Council on Legislation and on Scientific Affairs approved *White*'s lopsided statements: 'Difficulties arise in attempting to recommend a suitable range for dietary salt because of the tremendous range of biological tolerance in normal human beings, the widely different levels of salt appetite, and the cultural significance which salt has in relation to food.' He then went on to quote from a 1974 statement by the Committee on Nutrition of the American Academy of Pediatrics.

(a) 'The role of salt intake as an environmental factor in the induction of hypertension has still to be defined. For 80% of the population in this country, present salt intake has not been demonstrated to be harmful, i.e., hypertension has not developed...' Comment: this statement ignores 10% with moderately elevated blood pressure and 40% of the older population >60 years with hypertension.

(b) 'Salt appetite for some is an important expression of personal preference... in relation to diet; for others, salt-containing foods have important cultural values. Present evidence does not provide a firm basis for advising a change in the dietary salt intake for the general population.' Comment: *acquired* salt taste.

White [20] apparently had to change his mind because on March 30, 1979 he published a report: 'Nutrition: A Medical, Political and Public Issue' and wrote among others: 'It is likely that sodium chloride will soon be removed from the list of Generally Recognized as Safe (GRAS) food ingredients. The Select Committee on GRAS Substances Federation of American Societies for Experimental Biology, has expressed the opinion that salt intake should be reduced in the United States. Recommendations concerning sodium intake are also under consideration by the AMA Council on Scientific Affairs.'

The AMA position paper of 1977 is almost identical with an Editorial in the *Lancet* [4] which categorically stated: 'Current practice allows no place for salt restriction in the management of hypertension. The doctor who tells his hypertensive patient with normal renal function to avoid salt is wrong on two grounds. Firstly, the reduction in salt intake with such advice falls far short of that required to produce a significant effect upon blood pressure. Secondly, existing drugs are potent enough to render the misery of effective lifelong salt restriction unnecessary.' Fortunately, during the 5 years which have elapsed since this dogma was published, substantial evidence has accumulated to alter the attitude of physicians towards the dietary approach in the treatment of uncomplicated essential hypertension. Foremost, the classical review article by *Freis* [5] must be mentioned. His conclusion is as follows: 'Although the mechanism of essential hypertension is still obscure, the evidence is very good if not conclusive that reduction of salt in the diet to below 2 g/day would result in the prevention of essential hypertension and its disappearance as a major public health problem' When the pathologist, Dr. *Kaunitz* [quoted by *Meneely and Battarbee*, 15], who promoted the theory of a viral etiology of atherosclerosis, became involved with salt, he argued: 'Extra salt may be correlated with emotional stimulation... that it may be consciously or unconsciously pleasurable... as a causal factor in the craving for salt.' *Meneely and Battarbee* [15] dismissed these theories as simple speculations without any supporting evidence. They go on to criticize statements such as the one quoted from the *Lancet* [4] Editorial: 'For many patients (with essential hypertension) severe restrictions of sodium are not necessary... We are not locked into the high sodium-low potassium environment and we can do something about it. Human inertia and habit are formidable antagonists to dietary changes.'

Weight and Blood Pressure

The two major and unequivocal predictors of an individual's blood pressure are the blood pressure of first-degree relatives and the patient's weight. The mechanism of this correlation is not known. *Dahl* [3] felt that it was because fat people took in more salt with an increased number of calories. In support of this, he reported that the massively obese who were reduced on a low-calorie diet, but whose sodium intake was kept constant, did not have lower blood pressures at the end of the study. However, if the obese people were given a low-sodium diet without a change in weight, their blood pressure fell. There was a note in his study that the obese patients almost without fail, had a prompt drop in blood pressure with sodium restriction. *Dahl* stated that in his experience only one half of non-obese hypertensives responded similarly. Therefore, it is possible that obesity sensitizes the organism to the effect of salt. There are two studies of note on the effect of weight loss on blood pressure. In our Evans County hypertension intervention study we were able to lower blood pressure considerably through a combination of weight reduction, sodium restriction and potassium supplemention over a 12-month observation time. The fact that weight reduction alone is not the total explanation for the drop in blood pressure was made apparent by consideration of the association between weight reduction and blood pressure changes [7]:

Although there was a statistically significant association, the correlation coefficient is small (0.25) and therefore weight reduction *per se* statistically explained only a small proportion of all the blood pressure changes. Alternative explanations include the benefits of sodium restriction to less than 3 g/day and the possible hypotensive effect of the use of potassium supplements. In a recent report from Israel [18] blood pressure dropped with weight loss and, in the words of the authors of this report, 'independently of salt intake'. *Reisin et al.* [18] concluded: 'In a population of overweight hypertensive patients, up to 65% may reduce their blood pressure after a weight-reduction program with normal salt intake.' This erroneous conclusion was reached on the basis of a weight reduction program for 16 hypertensive men and only 8 women. The first group, by significantly reducing weight, must have followed the prescribed caloric restriction to 1,000 or less calories. It is quite clear that by lowering the caloric intake from such salt-rich nutrients as meat, milk products, bread and canned vegetables, to name but a few, the sodium content of such a diet is automatically reduced. In obese persons who lose weight, quite commonly a rather large loss of extracellular fluid volume occurs in the first month of weight reduction. This is, of course, associated with increased urinary sodium excretion. In the second group of patients, 42 men and 15 women, who were placed on an antihypertensive drug therapy *and* weight reduction, it is not surprising that large blood pressure reductions were demonstrated (37 mm Hg systolic, 23 mm Hg diastolic). Again, the influence of sodium depletion in this group, through sodium-eliminating diuretics, cannot be denied.

In our opinion, the classical experiments by *Dahl* [3], concluding that the fall in blood pressure with weight loss is due to the concomitant reduction in salt intake, have not been disproven. In addition, numerous observations, epidemiological as well as experimental in nature, have unequivocally demonstrated that high salt intake significantly contributes to the development of hypertensive blood pressure levels in individuals who may be genetically predisposed or may have a lower threshold for sodium intake and sodium disposal. Populations habitually living on very low sodium intakes do not show the characteristic rise in blood pressure with age seen in all industrialized nations, and hypertension is practically non-existent, again in sharp contrast to salt-eating societies.

At this time of increasing knowledge about the treatment of blood pressure elevation, it would be unfortunate if conclusions as quoted above would reach the lay press and add to the confusion of our patients. The goal of every physician treating hypertensives must be to (1) reduce weight (60% of our hypertensives are $\geq$20% above normal weight); (2) to lower the salt intake, and (3) to use antihypertensive drug therapy – in this order. Eventually, a large number of patients could either be placed on a reduced drug intake or the medication might be discontinued if the patient can be motivated to live with a low-salt diet.

A very intriguing experimental observation was reported by *Jung et al.* [9]. A hypotensive response to a low-calorie diet was explained by a fall in noradrenaline plasma concentrations and a marked reduction in urinary 4-hydroxy-3-methoxy mandelate. Overactivity of the sympathetic system has been implicated in the pathogenesis of some cases of hypertension with raised plasma concentrations of noradrenaline and adrenaline. The disadvantage of this experiment is two-fold,

(1) the 11 obese women were all normotensive, thus, it would be fascinating to repeat the observation with hypertensive individuals and extend the trial beyond 11 days;

(2) the sodium intake was severely restricted to 51 mEq/24 h. Therefore, one cannot be sure whether the significant blood pressure decrease cannot – again – be explained by the sodium deprivation. Nevertheless, it is important to look for other parameters of change under a weight reduction diet among obese hypertensives to elucidate the mechanism of blood pressure normalization.

Linoleic Acid Intake and Blood Pressure

Several experiments with animals fed on a linoleic-acid-enriched diet have convincingly demonstrated a saliuretic effect which in turn contributed to significant decreases in blood pressure in spite of continued high sodium intake. The results can be explained on the basis of increased prostaglandin production as most prostaglandins are known to cause increased sodium and water excretion. The background, design and results of these studies were recently reviewed by *Vergroesen* [in press, 1979].

The fact that a diet containing 10–15% calories from vegetable oils and

Table II. Blood pressure changes (mean ± SD) on a linoleic-acid-enriched diet [2]

Time	Systolic blood pressure mm Hg	Diastolic blood pressure mm Hg
6 months before	139 ± 5.2	97.5 ± 4.9
Pre-experimental	134.9 ± 10.0	92.4 ± 9.2
Week 1	133.1 ± 7.8	90.8 ± 2.5
Week 2	130.5 ± 7.5	88.3 ± 4.7
Week 3	133.5 ± 9.5	82.5 ± 9.6**
Week 4	127.0 ± 6.5	84.6 ± 7.4*
3 months off diet	136.1 ± 12.7	90.7 ± 6.8

* $p < 0.05$; ** $p < 0.01$ compared to pre-experimental value.

margarine has no adverse effects and the positive results from animal experiments conducted in East Germany and in the Netherlands persuaded us to join in two short-term human dietary trials. Both experiments did not prove beyond the shadow of a doubt that a linoleic-acid-enriched diet was responsible for a marked drop in blood pressure levels *in spite of weight maintenance and continued high sodium consumption* as is customary both in Georgia and in North Carolina. In the first experiment, 8 patients with essential hypertension served as their own controls [2]. Since 1960, their diastolic blood pressure levels have ranged between 90 and 100 mm Hg. Under a dietary regime which increased absorbed linoleic acid by 1.7%, the blood pressures decreased significantly from baseline and returned almost to baseline levels 3 months after termination of the study (table II).*

The second study (fig. 2, 3) on adolescents with elevated blood pressure ($\geq$90 mm Hg DBP) resulted in modest blood pressure decrements [19]. However, only for the normal-weight group of adolescents with elevated blood pressure did we achieve statistically significant decreases in blood pressure levels in comparison with a control group consuming the usual high-fat diet. Since the two groups were regrettably small, no major importance can be attached to this finding. It appears to us that in the *obese* youngsters or adults with elevated blood pressures, weight reduction and sodium restriction are of overriding importance – independent of the linoleic acid intake.

* Other details of the experimental results are presented in Table III.

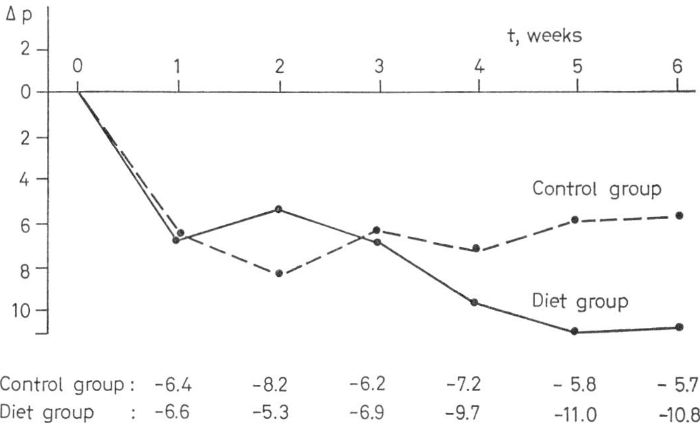

Fig. 2. Average systolic blood pressure decrease of the total group according to the baseline levels, divided into a control group (n = 18) and a diet group [n = 29; *Stern et al.,* in press, 1980].

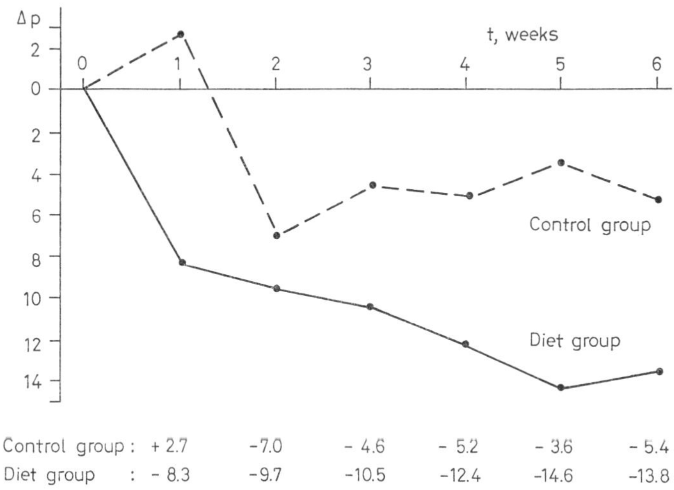

Fig. 3. Average systolic blood pressure decrease according to the baseline levels, divided into normal-weight controls (n = 5) and normal-weight subjects placed on a diet [n = 8; *Stern et al.,* in press, 1980].

The report by *Iacono et al.* [8] points in the same direction: This report summarizes two studies of age to evaluate the effect of diet on blood pressure and selected biochemical parameters.

Table III. Effect of increased dietary linoleate on serum and urine constituents [2]

Constituent	Pre-experimental serum, mean ±SD	Week 3, mean ±SD
Blood		
Cholesterol, mg/dl	219.6 ± 41.9	223.8 ± 40.3
Creatinine, mg/dl	1.09± 0.13	0.90± 0.22
Potassium, mmol/l	4.1 ± 0.35	4.2 ± 0.37
Sodium, mmol/l	137.3 ± 6.45	141.5 ± 3.34
Triglycerides, mg/dl	110.8 ± 33.1	121.5 ± 42.9
Urea nitrogen, mg/dl	15.3 ± 4.11	17.6 ± 6.21
Linoleate, cal.%	2.1 ± 0.7	3.8 ± 1.4*
Aggregation time, sec	149.5 ± 93.4	262.7 ±136.9
Urine		
Creatinine, mg/day	814.6 ± 212.8	2,184.3 ±665.8*
Potassium, mmol/day	52.6 ± 20.7	51.0 ± 19.7
Sodium, mmol/day	203.1 ± 85.1	185.5 ± 61.1
Urea nitrogen, g/day	12.81± 5.31	12.24± 3.87
Urine volume, ml	1,691 ±1,090	1,400 ±719
Creatine clearance, ml/min	52.68± 17.12	177.32± 65.60*
Urea clearance, ml/min	66.01± 29.82	56.82± 8.03

* $p < 0.01$ employing the two-tailed paired t test.

In the first study conducted in human subjects, 40–60 years of age *Iacono et al.* showed that feeding natural diets containing 25% fat calories with a P/S ratio of 1, for 40 days reduced significantly systolic and diastolic blood pressures in males and females after the 40-day feeding period.

The second study by *Iacono et al.* was conducted on hypertensive and normotensive groups fed natural diets for a 21-day stabilization period, followed by two 30-day experimental periods. 'The two groups were fed 25% fat calories with a P/S ratio of 1, or 44% fat calories with a P/S ratio of 0.3. Systolic and diastolic blood pressures were reduced to normal levels in the hypertensive group when the 25% fat calories, P/S ratio 1 were fed. When the same group was fed the 44% fat calories, P/S ratio 0.3, the blood pressure returned to almost the initial levels. Preliminary attempts to assess changes in urinary excretion of Na and K and total fluid volume suggested that these components were *excreted in higher amounts during the low-fat feeding period.*

Subjects in both studies consumed 8–10 g of NaCl, and maintained their body weight during the course of the experiments.

It is suggested that the reduction in blood pressure observed in these studies may have been due to the normalization of kidney function and was mediated through improved prostaglandin metabolism as a consequence of the higher levels of linoleic acid in the diet.

Finally, the first epidemiological evidence comes from a paper by *Oster et al.* [16]: 'In a natural, "free living population", of 800 men, surveyed with the purpose of health and nutrition assessment within an epidemiological design, the relationship of linoleic acid intake to other "risk factors" was examined. Two alternative methods of both short and long term linoleic acid intake assessment were incorporated into the study: a 24 hour dietary recall which was taken by one experienced dietician, and a fat biopsy for analysis by gas chromatography of relative fatty acid composition, as the C18:2 content of adipose tissue has been shown to reflect long term dietary intake of linoleic acid in men.

From 24 hour recall of linoleic acid intake estimations, the mean intake was calculated to be 18 g/day, and a positive correlation with urinary sodium concentration was seen (p<0.05).

With the measurement of relative composition of linoleic acid in fat, many more and stronger correlations were seen. Most interestingly was a strong negative relationship with both systolic and diastolic blood pressure (p<0.001).'

References

1   Addison, W.L.T.: The use of sodium chloride, sodium bromide and potassium bromide in cases of arterial hypertension which are amenable to potassium chloride. Can. med. Ass. J. *18:* 281 (1928).

2   Comberg, H.U.; Heyden, S.; Hames, C.G.; Vergroesen, A.J., and Fleischman, A.I.: Hypotensive effect of dietary prostaglandin precursor in hypertensive man. Prostaglandins *15:* 193 (1978).

3   Dahl, L.K.: Salt and hypertension. Am. J. clin. Nutr. *25:* 231 (1972).

4   Editorial: Lancet *June:* 1325–1326 (1975).

5   Freis, E.D.: Salt, volume and the prevention of hypertension. Circulation *53:* 589 (1976).

5a Grim, C.E.; Luft, F.C.; Miller, J.Z.; Menbely, G.R.; Battarbee, H.D.; Hames, C.G., and Dahl, L.K.: Racial differences in blood presence in Evans Country, Georgia, J. Chron. Dis. *33:* 87–94 (1980).

6   Grollman, A. and Grollman, E.F.: The teratogenic induction of hypertension. J. clin. Invest. *41:* 710–714 (1962).

7   Heyden, S.; Tyroler, H.A.; Hames, C.G.; Bartel, A.; Thompson, J.W.; Krishan, I., and Rosenthal, T.: Diet treatment of obese hypertensives. Clin. Sci. mol. Med. *45:* 209s (1973).

8   Iacono, J.M.; Marshall, M.W.; Machin, J.F.; Dougherty, R.M.; Weinland, B.T., and Canary, J.J.: Dietary fat and high blood pressure (personal comm., 1978).

9   Jung, R.T.; Shetty, P.S.; Barrand, M.; Callingham, B.A., and James, W.P.T.: Role of catecholamines in hypotensive response to dieting in normotensives. Br. med. J. *i:* 12–13 (1979).

10 Kempner, W.: Radical dietary treatment of hypertensive and arteriosclerotic vascular disease, heart and kidney disease, and vascular retinopathy. GP *9:* 71 (1954).

11 Langford, H.G. and Watson, R.L.: in Oglesby, Electrolytes and hypertension in epidemiology and control of hypertension, pp. 119–128 (Stratton, New York 1975).

12 Langford, H.G.; Watson, R.L., and Douglas, B.H.: Factors affecting blood pressure in population groups. Trans. Ass. Am. Physns *81:* 135 (1968).

13 Meneely, G.R. and Ball, C.O.T.: Experimental epidemiology of chronic sodium chloride toxicity and the protective effect of potassium chloride. Am. J. Med. *25:* 713–725 (1958).

14 Meneely, G.R.; Ball, C.O.T., and Youmans, J.B.: Chronic sodium chloride toxicity: protective effect of added potassium chloride. Ann. intern. Med. *47:* 263–273 (1957).

15 Meneely, G.R. and Battarbee, H.D.: Sodium and potassium. Nutr. Rev. *34:* 225 (1976); and Expert statement to the Select Committee on Nutrition and Human Needs, US Senate 1977.

16 Oster, P.; Arab, L.; Schellenberg, B.; Heuck, C.C.; Mordasini, R., and Schlierf, G.: Dietary linoleic acid and health status of Heidelberg men. Blood Pressure and Adipose Tissue Linoleic Acid. Res. Exp. Med. (Berl.) *175:* 287–291 (1979).

17 Ram, C.V.S. and Kaplan, N.M.: Diuretics and sodium restriction in the treatment of hypertension: effects on potassium wastage and blood pressure control. Abstract. Circulation *57/58:* suppl. II, p. 164 (1978).

18 Reisin, E.; Abel, R.; Modan, M.; Silverberg, D.S.; Eliahou, H.E., and Modan, B.: Effect of weight loss without salt restriction on the reduction of blood pressure in overweight hypertensive patients. New Engl. J. Med. *298:* 1–26 (1978).

19 Stern, B.; Heyden, S.; Miller, D.; Latham, G.; Klimas, A., and Pilkington, K.: Intervention study in high school students with elevated blood pressures. Nutr. Metab. *24:* 137–147 (1980).

20 White, Ph.: Nutrition: a medical, political and public issue. J. Am. med. Ass. *241:* 1407–1408 (1979).

Prof. S. Heyden, MD, PhD, Duke University Medical Center,
Department of Community Medicine, PO box 2914, Durham, NC 27710 (USA)

Discussion

Dr. Evans: We have heard good reasons for advising an increase in dietary linoleic acid intake but, in doing so, we risk criticism from some quarters that we may then be exposing people to increased risk of colorectal cancer. How should we counter this attack?

Prof. Heyden: I have answered this question comprehensively in an editorial for *Nutrition and Metabolism* [17: 321–328, 1974] under the title 'Polyunsaturated Fatty Acids and Colon Cancer'. The most frequently quoted study, alleged to show an association between cancer and polyunsaturated fatty acid is that by *Dayton* and his colleagues. In this long-term study, the dietary experimental group developed 31 cancers in contrast to the non-diet-adhering control group with only 17 cancers developing over the same time period of observation. However, I expressed my surprise that no one had commented on the high rate of non-adherence among men randomly assigned to the dietary experimental group. 10 of these men adhered less than 10% to the diet over an 8-year period and 2 additional men adhered only between 10 and 20% of the total time to this diet. Taking those 12 men out of the 31 cancer cases, the numbers of patients developing cancer control group and experimental group become almost equal.

The second study frequently quoted is that by *Rose* and his colleagues who showed, from a retrospective study of 90 cases of colon cancer pooled from several long-term studies, that low serum cholesterol concentrations would predict colon cancer development. However, Dr. *Rose* kindly provided me with a detailed table from which it turned out that 48 patients, i.e. more than 50% of the total number of colon cancer patients, died within 5 years after their last cholesterol determination. If we accept the experience of oncologists that colon cancer, in comparison to other more aggressive malignancies, is a relatively slow-progressing type of cancer, we can assume that people dying within 5 years of their last cholesterol determination must have had either regional involvement or widespread metastatic disease. I would like to suggest that, at least in the cases dying within the 5-year period, the cholesterol level would have had to be below expectation. 33 cases of colon cancer died within 1–3 years after their last cholesterol determination. These patients would most likely fit the common description of colon cancer patients dying with marked anemia, significant loss of body fat, protein, electrolyte abnormalities and increased basal metabolic rate as well as energy expenditure despite the reduced dietary intake – just the most obvious manifestations of a profound systemic derangement of the host metabolism. Dr. *Rose* added in a personal communication: 'that the cholesterol deviation to lower levels is greater when the interval from screening to diagnosis is short, i.e., in sick men. If we take out the men found within 1 year to have cancer, there is not in the remainder any correlation between cholesterol deviation and interval to diagnosis.'

Ederer and his colleagues published the combined experience of studies in Oslo, London, Helsinki and Faribault. The occurrence of cancer (total of diet and post diet phase) was 7.7% in the dietary group and 10.9% among the controls.

The Anticoronary Club in New York after 13 years with the Prudent Diet demonstrated a cancer incidence of 1.59% among 378 active dieters: an incidence of 1.75% among 853 inactive persons who had quit the diet and a cancer incidence of 1.88% among 533 controls. This again would be evidence against the allegation that polyunsaturated fatty acids may cause cancer.

Higginson found no differences in the regular use of vegetable or animal fat in a retrospective study on 340 patients with cancer of the colon and rectum and 1,200 controls. Dr. *Stamler*, at the recent American Health Foundation Meeting (April 1979) in New York made it quite clear from the Pooling Project Study in the USA, involving 6,000 men observed over a 10-year period, that there was an increased incidence of coronary heart mortality with every higher cholesterol subgroup. However, cancer mortality was the same in every cholesterol subgroup.

We agree with *Keys* who wrote in a personal communication: 'It still seems mysterious what the cholesterol-colon cancer data [of *Rose et al.*] reflect and mean. Associations that emerge from purely statistical analysis, with no basis of pathogenetic theory or support from experiment, may be the clue to other researches of importance, but by themselves may not have any meaning.'

Prof. Hartman: Much has been said today about the inverse relationship between HDL and coronary morbidity or risk. I would like to ask the epidemiologists: 'where would you place, at the present state of art, the HDL in the risk profile?'

What is the additional information provided by the fraction in excess of the total cholesterol in establishing a risk profile on the one hand in a population and on the other in an individual patient?

Prof. Heyden: Total cholesterol determination remains a good predictor of ischemic heart disease. We still have some difficulties in the laboratory in obtaining exact HDL determinations. Recent LRC studies have shown a surprising and very interesting close correlation between total cholesterol and LDL level: the higher the total cholesterol, the higher the LDL concentration.

Prof. Epstein: For the individual, at the clinical level, it is important and useful to determine the HDL. At the community level, cholesterol determination is quite adequate. Do not forget that about 15–20% of the men in our population have cholesterol levels over 260 and 20% of these men have enough HDL in the serum to protect them efficiently.

Nutr. Metab. *24* (Suppl. 1): 65–73 (1980)

Plasma Lipoproteins in Maturity Onset Diabetes

M. Mancini, A. Rivellese, P. Rubba and G. Riccardi

Center for Arteriosclerosis and Metabolic Diseases, Semeiotica Medica,
2nd Faculty of Medicine, Naples

Key Words. Lipoprotein metabolism · Hyperlipoproteinemia · Maturity onset diabetes ·
Plasma insulin · Obesity · Lipoprotein lipase · Intravenous fat tolerance test

Abstract. Plasma lipoprotein abnormalities in maturity onset diabetes (MOD) reflect
both enhanced production and impaired removal of triglyceride-rich lipoproteins. Hyperglycemia and hyperinsulinemia lead to overproduction of very-low-density lipoproteins
by the liver. Fat tolerance is reduced in MOD patients: this might be due to low lipoprotein
lipase activity (LLA) and/or to low incorporation of LLA-released fatty acids into adipose
tissue glyceride. The finding of abnormal low-density lipoprotein composition, with relative
enrichment in triglyceride, suggests remnant particle accumulation.

It has been known for many years that hyperlipidemia in maturity
onset diabetes (MOD) most often reflects hypertriglyceridemia, rather than
plasma cholesterol or phospholipid increase. This indicates that lipoprotein
abnormalities are confined to very-low-density lipoproteins (VLDL), the
carriers of endogenous triglyceride (TG) towards peripheral tissues. In fact,
the most common lipoprotein pattern in diabetes is type IV hyperlipoproteinemia [1].

We have analyzed plasma cholesterol [2] and TG [3] concentrations in
patients with uncontrolled MOD who have recently been examined at our
diabetic clinic.

Figure 1 shows the distribution of total plasma cholesterol in MOD
outpatients (33 males, 53 females). The distribution for male and female
patients overlapped so that a single curve has been plotted for both sexes.
It can be seen that about 20% of diabetic outpatients have plasma cholesterol
levels above our upper normal values (244 mg/dl). The 90th percentile of

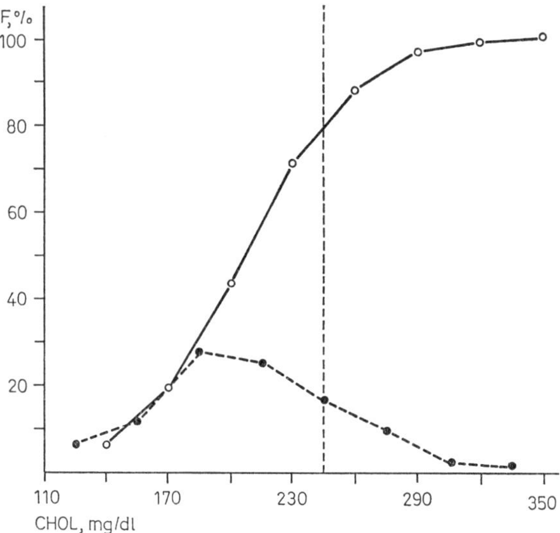

Fig. 1. Frequency (●) and cumulative distribution (○) of plasma cholesterol in patients with MOD (33 males, 53 females).

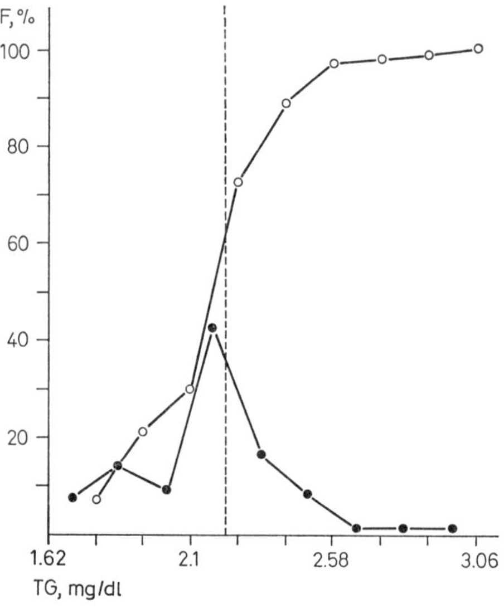

Fig. 2. Frequency (●) and cumulative distribution (○) of plasma TG (log) in patients with MOD (33 males, 53 females).

Table I. Types of hyperlipoproteinemia in diabetic patients (14 males, 15 females) and controls (25 males, 16 females)

Type	Diabetics		Controls	
	n	%	n	%
IIA	3	10	4	10
IIB	7	24	1	2
III	0	0	1	2
IV	5	18	4	10
Normal	14	48	31	76

cholesterol distribution in a random Neapolitan population sample after adjustment for sex and age was taken as the upper value. The individual values in this control material were adjusted for the age 45 years according to the criteria suggested by *Goldstein et al.* [4].

About 40% of the patients had plasma TG levels above our upper normal limits (mg 166/dl), calculated in the same way as for plasma cholesterol (fig. 2).

In a subsample of our patients (15 males, 14 females), preparative ultracentrifugation and quantitative analysis of cholesterol and TG in the individual lipoprotein classes were carried out. Thus accurate typing [5] could be performed with reference to the random population sample (table I). The most frequent types among diabetic patients were types IV and IIB, which are both characterized by elevated concentrations of VLDL (in type IIB also low-density lipoproteins, LDL, are increased).

Abnormalities leading to hyperlipidemia in MOD can be understood by examining three sites of plasma lipid transport [6]: (1) TG-rich lipoprotein secretion rate; (2) lipoprotein-lipase-mediated TG catabolism; and (3) remnant catabolism (fig. 3).

TG-Rich Lipoprotein Production

Input of TG-rich particles occurs from both exogenous and endogenous sources, leading to the synthesis of chylomicron in the intestine and of VLDL in the liver. During and after meals, free fatty acids enter the liver, where they may be esterified with glycerol to form TG. *De novo* synthesis of fatty acids also occurs in the liver, where fatty acids are synthetized from

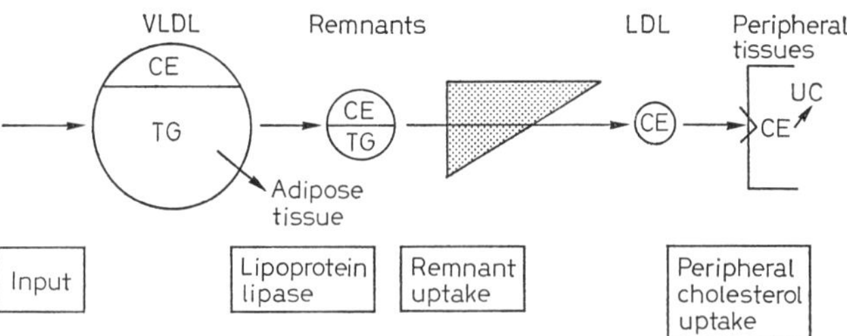

Fig. 3. Sites of plasma lipoprotein transport. CE = Cholesterol ester; UC = unesterified cholesterol [from *Brunzell et al.*, 6].

glucose or acetate [7]. Both of these processes are markedly influenced by changes in insulin and glucagon levels occurring with food intake.

Fatty acids in the cytosol of liver cells can either enter mitochondria, where oxidation takes place, or remain in the cytosol where they are esterified to TG. The fate of hepatocellular fatty acids also appears to be under hormonal control: glucagon enhances and insulin prevents mitochondrial fatty acid uptake and subsequent oxidation [8]. TG produced by the liver, together with cholesterol ester, is combined with the lipoprotein monolayer composed of phospholipid, unesterified cholesterol and apoprotein B, and secreted into the hepatic venous outflow as endogenous TG-rich VLDL. Thus, in healthy subjects with normal body weights, plasma insulin concentration [9] is directly correlated to VLDL TG concentration (fig. 4).

In MOD, hyperglycemia and hyperinsulinemia are likely to lead to overproduction of VLDL [10], perhaps through effects on the sites of regulation of fatty acid synthesis and esterification in the liver. Enhanced fat-mobilizing lipolysis in adipose tissue might also contribute to the overproduction of VLDL by providing increased amounts of precursors (fatty acids) for TG synthesis in the liver [11].

In vivo studies on plasma TG turnover in MOD support this view [12]. However, it is difficult to separate in these patients, the individual effects of diabetes and obesity on the basis of TG turnover. Studies using radiopalmitate infusion have shown that the turnover rate of plasma VLDL TG is increased in nondiabetic obese patients, which accounts for their high plasma TG concentrations [13]. Obesity might thus explain the increased VLDL production in MOD; other studies, however, have demonstrated that TG increase is more marked in obese diabetics as compared to nondiabetic obese patients, for the same degree of overweight [14].

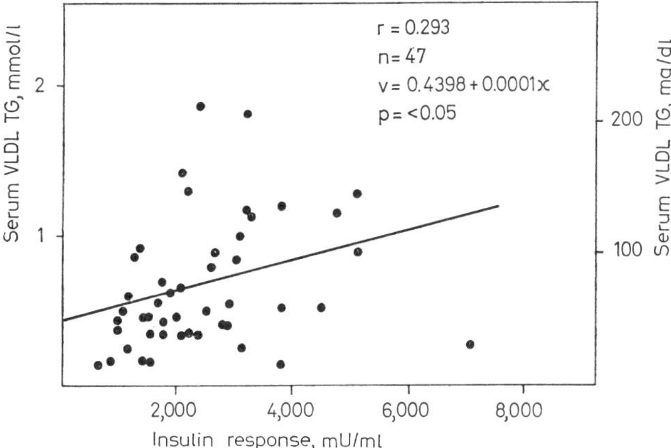

Fig. 4. Linear regression of serum VLDL TG concentration upon the sum of plasma insulin concentrations during an intravenous glucose load (0.5 g/kg).

Lipoprotein-Lipase-Mediated TG Catabolism

Secreted TG entering plasma in chylomicrons and VLDL are primarly transported to adipose tissue and muscle for storage and utilization. Lipoprotein lipase is the enzyme in adipose tissue that catalyzes TG uptake [15]. It is synthetized in adipocytes and, following secretion and transport to the capillary endothelial cell by unknown mechanisms, hydrolyzes lipoprotein TG at that site [16].

At least two of the three fatty acids, potentially releasable through TG hydrolysis, are then transported to the fat cell where they are reesterified with glycerol and stored as intracellular adipocyte TG (the fat droplet). The vast majority of TG in the human adipocytes is accumulated by this mechanism, little *de novo* extrahepatic lipogenesis from glucose occurring in man[17].

Impairment in the removal of plasma TG by peripheral tissues (adipose tissue, skeletal muscle, heart) has been demonstrated in MOD by an intravenous fat tolerance test [18], using a synthetic TG emulsion (Intralipid). The rate of disappearance from blood of intravenously injected Intralipid is a measure of fractional removal rate of both classes of TG-rich lipoproteins, chylomicrons and VLDL [19]. Slower Intralipid disappearance has been demonstrated in MOD. In addition, an inverse correlation has been shown (fig. 5) in untreated diabetics, between plasma TG concentration and Intra-

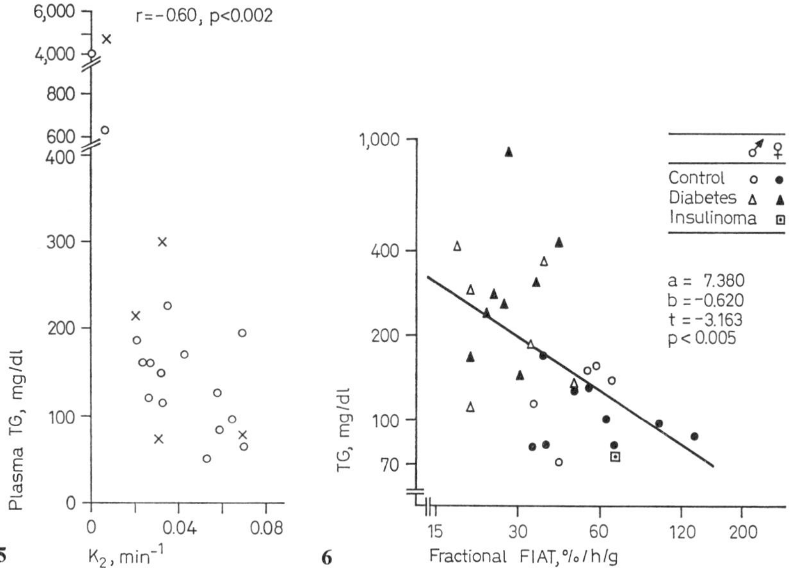

Fig. 5. Correlation between plasma TG concentration and fractional turnover of exogenous TG (K$_2$) in MOD.

Fig. 6. Linear regression of plasma TG (log) upon fractional FIAT (log) in control subjects, diabetics and an insulinoma patient.

lipid removal: thus it is likely that impaired lipoprotein TG removal contributes, at least in part, to hypertriglyceridemia in MOD [20].

Defective plasma TG removal by peripheral tissues, such as adipose tissue, might be determined by either or both of these mechanisms: (1) a low lipoprotein lipase activity (LLA) which impairs the hydrolysis of lipoprotein TG to free glycerol and fatty acids; (2) a defective incorporation of the LLA-released fatty acids into adipose tissue.

In MOD, LLA of adipose tissue is generally within the normal range [21, 22]; however, somewhat lower values are found in patients with elevated plasma TG concentration. As studies on LLA do not demonstrate a clear-cut defect in MOD, it is interesting to evaluate also the process of fatty acid incorporation into adipose tissue (FIAT). FIAT has been demonstrated to be often impaired in primary hypertriglyceridemia [23–25] and it has been proposed as a contributing factor for the removal defect which is often associated with high plasma TG concentrations. We have found an inverse correlation between plasma TG concentration and FIAT in a group of 29

Table II. Lipoprotein composition (mg/dl) in diabetic patients (14 males, 15 females) and controls (26 males and 16 females)

		VLDL[a]		LDL		HDL	
		TG	CHOL	TG	CHOL	TG	CHOL
Male	diabetics	65	15	40 ± 4***	134 ± 15	11 ± 2	43 ± 3
	controls	68	20	19 ± 1	116 ± 6	9 ± 1	46 ± 2
Female	diabetics	74*	15	33 ± 5**	134 ± 10	14 ± 2	46 ± 3*
	controls	39	12	18 ± 1	120 ± 7	12 ± 1	59 ± 4

* $p < 0.025$; ** $p < 0.01$; *** $p < 0.001$.   [a] Geometric means

subjects including 14 normoglycemic-normolipidemic controls, 14 MOD patients and 1 patient with insulinoma and hyperinsulinism (fig. 6), These findings suggest that low FIAT might also contribute to the production of hypertriglyceridemia in MOD [26].

Remnant Lipoprotein Catabolism

Following hydrolysis and removal of the TG moiety from chylomicrons and VLDL, with simultaneous removal of other lipid components, a lipoprotein is formed which has been termed 'remnant'. The intermediate-density fraction or LDL_1 (d = 1.006–1.019) largely consists of remnant particles, but the remnants derived from chylomicrons and large endogenously synthetized VLDL also appear to be distributed in the VLDL range of density. Once formed, the remnant has a short half-life in plasma and is probably taken up by the liver [6].

Very few information is available on concentration and turnover of remnant particles in MOD. In our material, we made an interesting finding on this point (table II). When we compare the mean concentration of cholesterol and TG in the lipoproteins of diabetics with those of a healthy control group of matched ages (40–50 years), we see that LDL are abnormal. In particular, this lipoprotein class shows a relative enrichment in TG in spite of normal cholesterol content. This would suggest a relative increase in the intermediate-density subfraction (LDL_1).

This is an agreement with a previous finding [1] showing a relative enrichment of LDL in their TG content in diabetic patients. These data suggest that abnormalities in remnant lipoprotein catabolism in MOD cannot be excluded. Further studies on lipoprotein turnover in diabetic patients, particularly those with normal lipoprotein levels, are needed, however, to test the validity of this hypothesis.

References

1 Schönfeld, G.; Birge, C.; Miller, J.P.; Kessler, G., and Santiago, J.: Apolipoprotein B levels and altered lipoprotein composition in diabetes. Diabetes *23:* 827 (1974).

2 Abel, L.L.; Levy, G.B.; Brodie, B.B., and Kendall, F.E.: A simplified method for the estimation of total cholesterol in serum and demonstration of its specificity. J. biol. Chem. *195:* 357 (1952).

3 Pantuly, G.V.A.; Anderson, J.T., and Keys, A.: A rapid and specific method for serum triglyceride determination. Fed. Proc. *24:* 438 (1964).

4 Goldstein, J.; Hazzard, W.R.; Schrott, H.G.; Bierman, E.L., and Motulsky, A.G.: Hyperlipidemia in coronary heart disease. I. Lipid levels in 500 survivors of myocardial infarction. J. clin. Invest. *52:* 1533 (1973).

5 Rubba, P.; Farinaro, E.; Postiglione, A.; Oriente, P., and Mancini, M.: Multiple lipoprotein abnormalities in primary hyperlipidemia in Naples; in Carlson, Paoletti, Sirtori and Weber, Int. Conf. on Atherosclerosis, p. 117 (Raven Press, New York 1978).

6 Brunzell, J.D.; Chait, A., and Bierman, E.L.: Pathophysiology of lipoprotein transport. Metabolism *27:* 1109 (1978).

7 Barter, P.J.; Nestel, P.J., and Carrol, K.F.: Precursor of plasma triglyceride fatty acid in humans. Effects of glucose consumption, clofibrate administration and alcoholic fatty liver. Metabolism *21:* 117 (1972).

8 Mc Farry , J.D. and Foster, D.W.: Ketogenesis and its regulation. Am. J. Med. *61:* 9 (1976).

9 Hales, C. and Randle, P.: Immunoassay of insulin with insulin-antibody precipitate. Biochem. J. *88:* 137 (1963).

10 Olefsky, J.M.; Farquhar, J.W., and Reaven, G.M.: Reappraisal of the role of insulin in hypertriglyceridemia. Am. J. Med. *57:* 551 (1974).

11 Bjorntorp, P. and Hood, B.: Studies on adipose tissue from obese patients with or without diabetes mellitus. I. Release of glycerol and free fatty acids. Acta med. scand. *179:* 221 (1966).

12 Nikkila, E.A. and Kekki, M.: Plasma triglyceride transport kinetics in diabetes mellitus. Metabolism *12:* 1 (1973).

13 Nestel, P.J. and White, H.M.: Plasma free fatty acid and triglyceride turnover in obesity. Metabolism *17:* 1122 (1968).

14 Nikkila, E.A.: Triglyceride metabolism in diabetes mellitus. Progr. biochem. Pharmacol., vol. 8, p. 271 (Karger, Basel 1973).

15 Fielding, C.J. and Havel, R.J.: Lipoprotein lipase. Archs Path. *101:* 255 (1977).

16 Blanchette-Mackie, E.J. and Scow, R.O.: Sites of lipoprotein lipase activity in adipose tissue perfused with chylomicrons. Electron microscopic cytochemical study. J. Cell Biol. *51:* 1 (1971).

17 Patel, M.C.; Owen, O.E., and Goldman, L.L.: Fatty acid synthesis by human adipose tissue. Metabolism *24:* 161 (1975).

18 Lewis, B.; Boberg, J.; Mancini, M., and Carlson, L.A.: Determination of the intravenous fat tolerance test with Intralipid by nephelometry. Athersoclerosis *15:* 83–86 (1972).

19 Rossner, S.: Studies on an intravenous fat tolerance test. Methodological, experimental and clinical experience with Intralipid®. Acta med. scand., Suppl. 564, p. 1 (1974).

20 Lewis, B.; Mancini, M.; Mattock, M.; Chait, A., and Russel Fraser, T.: Plasma triglyceride and fatty acid metabolism in diabetes mellitus. Eur. J. clin. Invest *2:* 445 (1972).

21 Persson, B.: Lipoprotein lipase activity of human adipose tissue in health and in some diseases with hyperlipidemia as a common feature. Acta med. scand. *193:* 457 (1973).

22 Nikkila, E.A.; Huttunem, J.K., and Enholm, C.: Relationship to plasma triglyceride of post-heparin plasma lipoprotein lipase and hepatic lipase in diabetes mellitus. Diabetes *26:* 11 (1977).

23 Walldius, G.; Olsson, A.G.; Rubba, P., and Carlson, L.A.: Metabolic defects in adipose tissue in hypertriglyceridemia: new aspects on the pathogenesis of hypertriglyceridemia; in Vague and Boyer, The regulation of the adipose tissue mass, p. 327 (Excerpta Medica, Amsterdam 1973).

24 Carlson, L.A. and Waldius, G.: Fatty acid incorporation into human adipose tissue in hypertriglyceridemia. Eur. J. clin. Invest. *6:* 195 (1976).

25 Rubba, P.: Fractional fatty acid incorporation into human adipose tissue (FIAT) in hypertriglyceridemia. Atherosclerosis *29:* 39 (1978).

26 Rubba, P.; Pezzella, G.; Postiglione, A., and Mancini, M.: Fatty acid incorporation into human adipose tissue in maturity onset diabetes and a patient with an insulinoma; in Crepaldi, Lefebvre and Alberti, Diabetes, obesity and hyperlipidemias, p. 223 (Academic Press, London 1978).

M. Mancini, MD, Università degli Studi di Napoli, Istituto di Semeiotica Medica II, Policlinico, Via Pansini, 5, I-80131 Napoli (Italy)

Nutr. Metab. *24* (Suppl. 1): 74–89 (1980)

Dietary Effects on Certain Adrenal Cortical Functions in the Rat

G.S.Boyd, A.M.S.Gorban and Margaret E.Lawson

Edinburgh University Medical School, Department of Biochemistry, Edinburgh

Key Words. Adrenal cortex · Cholesterol esterase · Cholesterol esters · Dietary fat · Cytochrome P_{450}

Abstract. The rate-limiting step in adrenal steroidogenesis is associated with the mitochondrial-cytochrome-P_{450scc}-dependent production of pregnenolone from cholesterol. This sterol side-chain cleavage reaction is influenced by the supply of cholesterol to the mitochondria. Cholesterol is stored as cholesterol esters while the cytosol contains a hormone-sensitive cholesterol ester hydrolase. This enzyme is activated by phosphorylation involving a cyclic AMP-dependent protein kinase and ATP; this enzyme preferentially attacks cholesterol oleate or cholesterol linoleate. The lipid composition of the adrenal cortex is influenced by diet so that animals on a low-fat diet tend to store cholesterol oleate and as the linoleate content of the diet is increased, the cholesterol linoleate content of the adrenal cortex increases. Animals maintained on a high erucate diet tend to store large amounts of cholesterol erucate in the adrenal cortex; such animals have an impaired adrenal cortical function. Animals maintained on a low-fat diet (marginally deficient in essential fatty acids), a linoleate-replete diet or a moderate erucate diet, all exhibited normal responses to ACTH and normal corticosterone production rates.

Structure and Function of the Adrenal Cortex

The adrenal glands lie closely applied to the upper aspects of the kidneys. Each gland consists of a central medulla and a surrounding adrenal cortex. Although in mammals these two components of the adrenal gland are fused together, they are nevertheless embryologically distinct. In this account, attention will be paid only to the adrenal cortex, which is composed of large polyhedral cells rich in lipids. The cells of the adrenal cortex are often classified into three zones. Under the fibrous capsule there lie the cells of

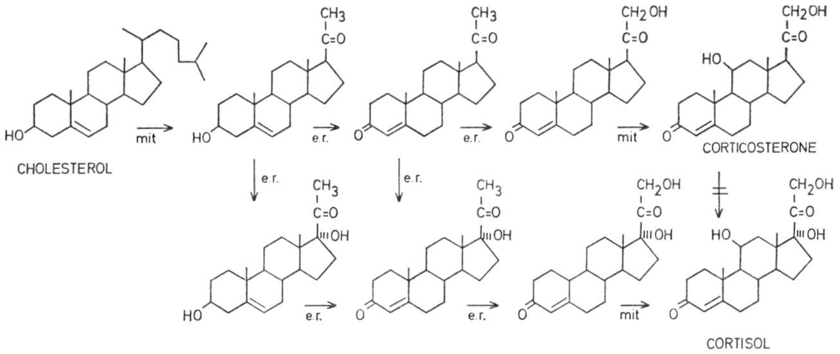

Fig. 1. Metabolic pathway for the conversion of cholesterol to corticosteroids in the adrenal cortex. mit = Mitochondria; e.r. = endoplasmic reticulum.

the zona glomerulosa and these make contact with the cells of the zona fasciculata which form the principal mass of the adrenal cortical cells. Under the zona fasciculata lie the cells of the zona reticularis containing compact cells which are poor in lipid. This description of the adrenal cortex applies only to the adult; in the fetus, zonation of the adrenal cortex is different [1]. Furthermore, although this description of the zones of the adrenal cortex applies to the rat, rabbit and man, there are other species in which this type of zonation does not appear to occur. There are many excellent reviews of the structure of the adrenal cortex [2–4].

Bilateral adrenalectomy leads to a complex series of metabolic changes such as increased urinary sodium loss, decreased blood pressure, decreased blood glucose, decreased mobilisation of tissue lipids and general muscular weakness [5]. Loss of adrenal cortical function is therefore virtually incompatible with survival in a normal environment. During the past 50 years there has been much research into the nature of the secretions of the adrenal cortex and the control of the secretions of this organ. Many steroids have been isolated from adrenal tissue, and from an analysis of the potency of these compounds in various biological assays a rational division of adrenal steroids into two distinct categories has evolved. This classification of the adrenal steroids into glucocorticoids and mineralocorticoids implies that the former appear to influence selectively carbohydrate and protein metabolism while the latter have a major influence on electrolyte balance. In man, the principal glucocorticoid is cortisol, and in the rat, the principal glucocorticoid is corticosterone. The structures of these hormones are shown in figure 1.

Control of the Adrenal Cortex

Since the adrenal cortex is the source of diverse steroid hormones which have powerful effects on metabolic processes, it is of importance to consider the factors which control the adrenal cortex. Adrenocorticotropin (ACTH) produced by the basophil cells of the anterior pituitary is a polypeptide of 39 amino acids of which the 24 amino acids from the N-terminal end of the molecule contain the active fragment of the hormone [6, 7]. The principal target tissues for ACTH are the adrenal cortex and adipose tissue [8, 9].

In the adrenal cortex, ACTH has its main effect on the two inner zones of the adrenal cortex, the zona fasciculata and the zona reticularis [10, 11], while the zona glomerulosa is controlled very largely by the pressor agent, angiotensin II [12, 13].

The secretion of ACTH in the rat and in man is known to be subject to a diurnal rhythm, and in addition there is also a pulsatile release of ACTH. In man, the plasma concentration of ACTH is low in the late evening and high in the early morning. As ACTH has a very rapid effect on the rate of secretion of adrenal steroids, the plasma concentrations of the adrenal corticoids closely follow the plasma concentrations of ACTH [14, 15].

Before considering how other factors such as diet may influence the adrenal cortex, it is essential to examine how the adrenal cortex appears to respond to ACTH with a rapid increase in output of steroid hormones. There is evidence to support the concept that the immediate precursor of the steroid hormones is intracellular cholesterol. In the adrenal cortex, cholesterol occurs largely in the form of cholesterol esters, and the quantity of esterified cholesterol in the adrenal cortex in some species may exceed 5% on a wet weight basis [16]. ACTH attaches to the plasma membrane of the adrenal cell; the hormone does not penetrate the cell [17] but activates a plasma membrane-bound adenyl cyclase [18]. As a result, there is an almost immediate increase in the intra-adrenal cell cyclic AMP (cAMP) concentration. The concentration of ACTH in the medium bathing the cells is related to the intracellular cAMP concentration which in turn is related to the increased output of corticosterone [18, 19].

cAMP and Steroidogenesis

The position of cAMP as a secondary messenger in the adrenal seems well established [20]. When adrenal cells *in vitro* are treated with cAMP (or an ester of cAMP), then these cells respond by an increased output of corticosterone in the absence of ACTH [21]. This suggests that ACTH produces a change in intracellular cAMP and the resulting elevated steroidogenic response is mediated by cAMP.

In adrenal tissue, cAMP is specifically bound to a protein in the adrenal cytosol and, as a consequence of this cAMP-protein association, there is activation of a cytosolic protein kinase [22, 23]. In the adrenal cytosol, there is a protein kinase composed of two subunits; one of these subunits is termed the 'receptor' subunit and the other the 'catalytic' subunit. When the two subunits are together, the protein kinase is in an inactive or restrained state. When cAMP binds to the receptor subunit, this causes this subunit to dissociate from the catalytic subunit so that the protein kinase activity is fully expressed. There are a number of reviews on this topic [24–26]. There remains the requirement to investigate which proteins involved in steroidogenesis in the adrenal cell are phosphorylated by the cAMP-dependent protein kinase.

The biosynthetic route between cholesterol and corticosterone or cortisol is well established. There is good evidence on the localisation and nature of the enzymes involved in the biosynthesis of the adrenal steroids. The main reactions are shown in figure 1. It will be noted in this scheme that various intermediates are transformed in different organelles so that, during the biosynthesis of the steroid hormones, there is a constant shuttling of intermediates into and out of certain organelles.

In many biosynthetic processes in which there are multiple steps, the first step is often the rate-limiting or controlling step. In steroidogenesis in the adrenal cortex, the first step is the transformation of cholesterol to pregnenolone in mitochondria, the cholesterol side-chain cleavage process [27–30]. This reaction has been examined in detail in attempts to establish whether ACTH directly or indirectly influences the enzymic reaction. The split of the cholesterol molecule between C_{20} and C_{22} is effected by a mitochondrial mixed-function oxidase located in the inner cristae of the mitochondria. The reaction requires O_2 and NADPH and involves sequential hydroxylation reactions. This mitochondrial mixed-function oxidase enzyme complex contains the haemoprotein cytochrome $P450_{scc}$ [31, 32]. The mitochondria of the adrenal cortex contain the conventional enzymes associated with mitochondrial function such as the enzymes of oxidative phosphorylation. In addition in the inner cristae of the adrenal cortex mitochondria there are at least three distinct forms of cytochrome P450 catalysing different hydroxylation events. There is the cholesterol side-chain cleavage complex, the steroid 11β-hydroxylase and the steroid 18-hydroxylase enzymes. Cytochrome P450 involved

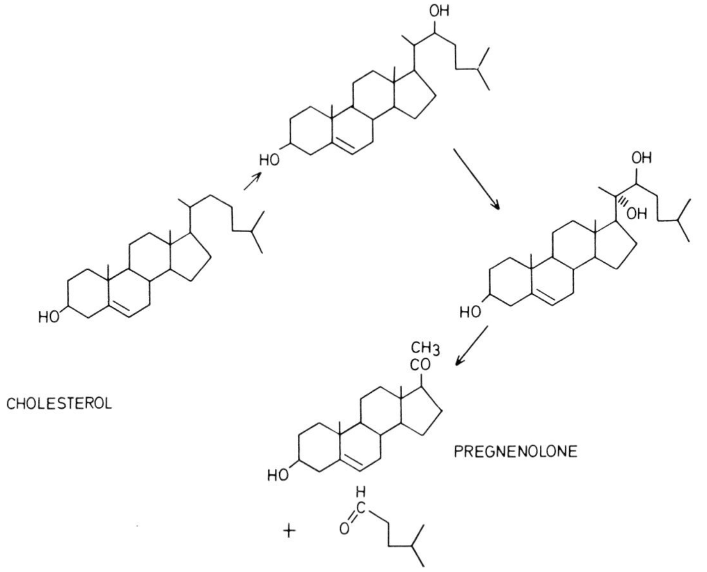

Fig. 2. The probable route for the conversion of cholesterol to pregnenolone in adrenal cortex mitochondria. Cholesterol is converted into (22R) 22-hydroxycholesterol and this in turn is converted into (20 R, 22R) 20,22-dihydroxycholesterol. The ultimate product of the cholesterol side-chain cleavage reaction is pregnenolone.

in the cholesterol side-chain cleavage reaction is often referred to as cytochrome $P450_{scc}$ while cytochrome P450 involved in the 11β-hydroxylase reaction is often termed cytochrome $P450_{11\beta}$. These mitochondria have an electron transport chain utilising NADPH as the electron donor to a specific flavoprotein and an iron sulphur protein finally donating electrons to one of the cytochromes P450 as follows:

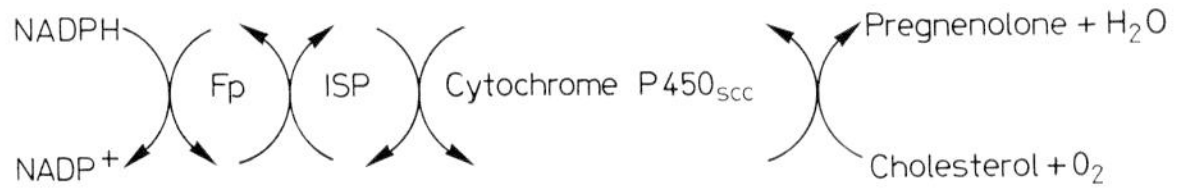

The specificity of each hydroxylase resides in the cytochrome P450 component of the multienzyme complex. It has been possible to purify partially the cytochrome P450 associated with the cholesterol side-chain cleavage reaction, and to reconstitute the electron transport chain of the NADPH, the flavoprotein and the iron sulphur protein with this cytochrome $P450_{scc}$. By the addition of substrate cholesterol to the cytochrome $P450_{scc}$ and the controlled supply of electrons and O_2 to the reaction mixture, the sequential steps in the conversion of cholesterol to pregnenolone have been elucidated [30]. These studies have confirmed that the hydroxylation reactions follow the pattern cholesterol – 22-hydroxycholesterol – 20, 22-dihydroxycholesterol – pregnenolone [33], as shown in figure 2.

Supply of Cholesterol to Adrenal Mitochondria

In the kinetic analysis of the individual steps in the cholesterol side-chain cleavage reaction, the cholesterol substrate must be in an unesterified state to combine with the cytochrome $P450_{scc}$. A slow step in the side-chain cleavage sequence appears to be the association of cholesterol with cytochrome $P450_{scc}$. The free cholesterol content of most cells and tissues appears to be quite delicately controlled, whereas the esterified cholesterol content can fluctuate markedly. Thus the free cholesterol present in the inner cristae of the adrenal mitochondria in the rat is only sufficient to last a few minutes under conditions in which the adrenal cells are required to produce adrenal steroids at the maximal rate [34, 35]. If it is accepted that the rate-limiting event in adrenal steroidogenesis is pregnenolone production in mitochondria, then in the rat adrenal cortex the mitochondrial cholesterol available for steroidogenesis would be limiting in a few minutes. If NADPH and O_2 supplies to the adrenal mitochondria are not limiting, then the rate of production of pregnenolone by these organelles is influenced by the availability of the substrate cholesterol [36]. If cholesterol is delivered to adrenal mitochondria in a small amount of a solvent such as acetone, or in a phospholipid:cholesterol micellar suspension, or in the form of a tissue lipoprotein isolated from the adrenal cell cytosol, there is an effective increment in the rate of mitochondrial production of pregnenolone from cholesterol. As shown in figure 3, the production rate of pregnenolone is stimulated in the presence of such cholesterol-bearing complexes. It is necessary to examine the source of cholesterol available to the adrenal mitochondria from which the mitochondrial cholesterol reserves may be supplemented during phases of maximal steroidogenic response.

Such a pool of non-esterified cholesterol available for steroidogenesis could be derived from various sources, as shown in figure 4. (i) The adrenal cell can synthesise cholesterol *de novo,* but whether the rate of cholesterol synthesis could be activated quickly enough to meet the demands of adrenal steroidogenesis has not been established. (ii) The mitochondrial pool of non-

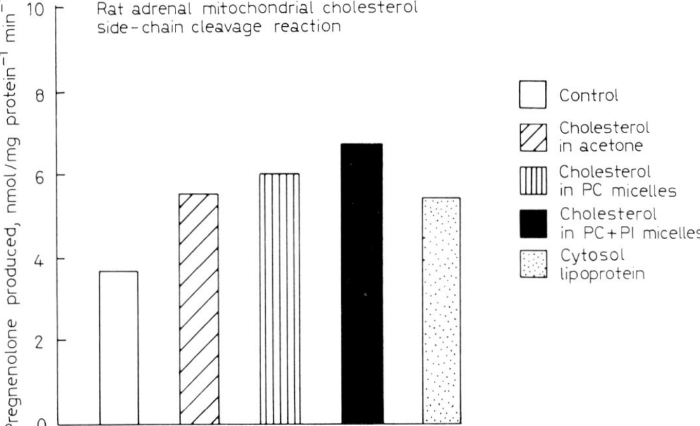

Fig. 3. Rat adrenal mitochondria isolated in the usual way [36] were pre-incubated for 15 min with cholesterol presented to the mitochondria as follows. (a) Control incubation without cholesterol supplement; (b) cholesterol added in 50 µl acetone to give a final concentration of 200 µM cholesterol; (c) cholesterol added as cholesterol-phosphatidyl choline micelles to give a final concentration of 200 µM; (d) cholesterol added in phosphatidyl-choline and phosphatidyl inositol micelles to give a final concentration of 200 µM cholesterol, and (e) cholesterol containing lipoprotein from the adrenal cytosol added again to give a final concentration of 200 µM cholesterol. In each case, the mitochondria were incubated at 37 °C, aerobically and the reaction was started by the addition of iso-citrate. The reaction rates were linear for 10 min.

esterified cholesterol could be supplemented by exchange reactions from non-esterified cholesterol in other membrane systems such as the plasma membrane or the endoplasmic reticulum. (iii) Free cholesterol could be obtained from the plasma lipoproteins, e.g. HDL and LDL by phagocytosis and lysosomal action. (iv) The controlled hydrolysis of esterified cholesterol present in the cytoplasm as lipid droplets could supply free cholesterol to the cell [34].

Hormone-Sensitive Cholesterol Ester Hydrolase

The lipid droplets in the adrenal cortex are rich in cholesterol esters and triglycerides. The adrenal cell contains a number of cholesterol ester hydrolases in lysosomes, mitochondria and in the cell cytosol. Situations which are known to increase the concentration of ACTH in blood plasma have been

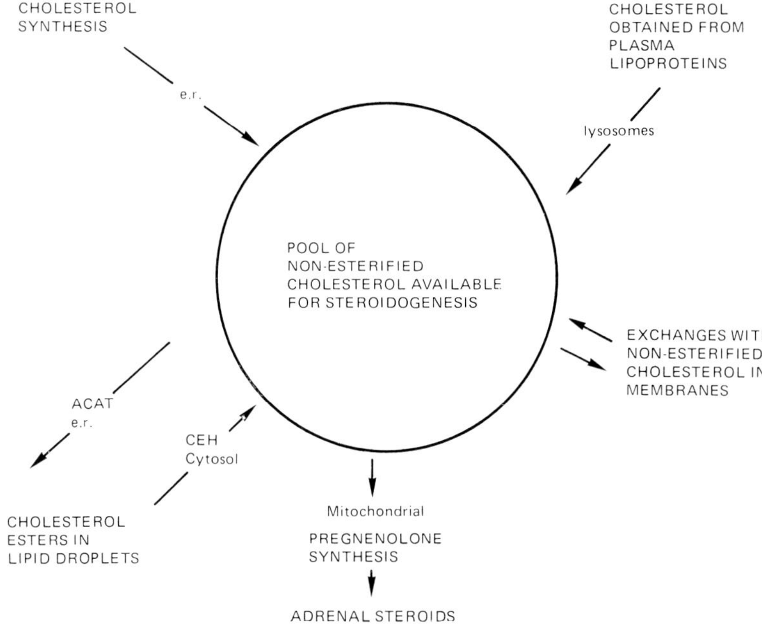

Fig. 4. Possible sources of free or non-esterified cholesterol in the adrenal cortex. e.r. = Endoplasmic reticulum; CEH = cholesterol ester hydrolase; ACAT = acyl CoA-cholesterol acyl transferase.

shown to produce a marked depletion of cholesterol esters within the lipid droplets isolated from rat adrenal glands. This has been interpreted as the result of an increased hydrolysis of cholesterol esters induced by stimulation of the adrenal cortex by ACTH.

In the adrenals of stressed rats, the activity of the cytosolic cholesterol ester hydrolase is increased [37], the concentration of cAMP is increased and the activity of a protein kinase is increased [37]. Rat adrenal cytosolic cholesterol ester hydrolase can be activated *in vitro* by the addition of cAMP and ATP in the presence of endogenous protein kinase [37–40]. This activation is associated with phosphorylation of the inactive enzyme, and the phosphorylation is achieved by the transfer of the γ-phosphate of ATP onto the inactive enzyme [38]. Conversely, it has been shown that there is present in the adrenal cytosol a magnesium-ion-dependent phosphoprotein phosphatase capable of hydrolysing the phosphate esterified to the active form of the cholesterol ester hydrolase, thus causing inactivation of the enzyme. The

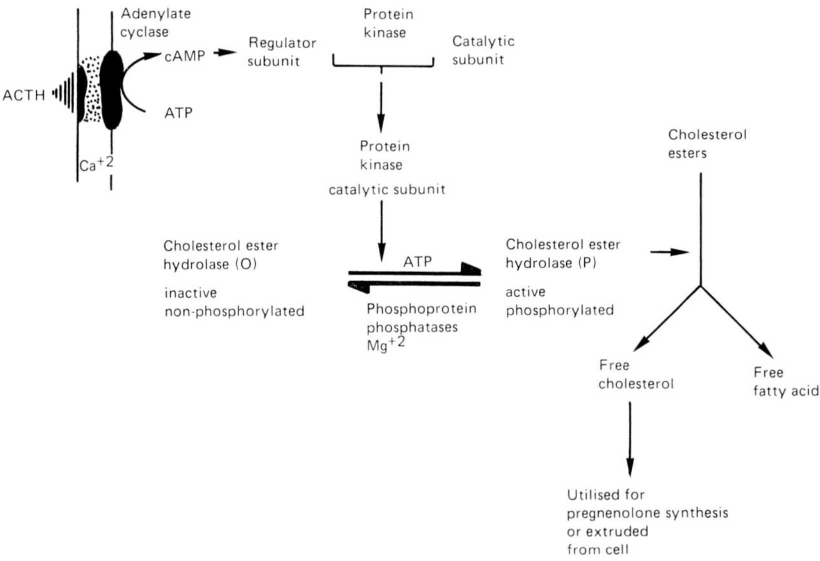

Fig. 5. Cascade reaction in the adrenal cortex cytosol initiated by activation of adenyl cyclase, elevated cAMP concentrations, and activation of the cytosolic cholesterol ester hydrolase.

control of the hydrolysis of cholesterol esters in the adrenal cortex may be under the influence of two opposing reactions associated with the covalent modification of the enzyme cholesterol ester hydrolase. This is shown in figure 5.

This may not be the only mechanism by which ACTH stimulates steroidogenesis. When animals or isolated adrenal cells are subjected to pre-treatment with a protein synthesis blocker, followed later by the administration of ACTH, then the ability of the adrenal to produce steroid hormones is greatly impaired [41–43]. However, cycloheximide pre-treatment does not block the cascade reaction that leads to the increase in intracellular free cholesterol [23]. Consequently, it has been argued that ACTH must participate in other processes also, for example the production of some 'labile protein' or 'peptide' involved in the intramitochondrial transport of cholesterol to the inner cristae for utilisation in the cholesterol side-chain cleavage reaction [23]. However, there is no evidence at present that changes in the lipid composition of the diet can affect the production of this labile factor in adrenal mitochondria.

The adrenal cortex contains substantial amounts of triglycerides as well as cholesterol esters. It has been shown that when the plasma ACTH concentration is raised, the triacylglycerol lipase activity of the adrenal cortex is also activated [44]. The mechanism for the activation of this enzyme appears to be similar to the mechanism of activation of the cholesterol ester hydrolase. This has resulted in investigations into the possibility that in the adrenal cortex there may be two independent enzymes, one operating on the cholesterol esters and the other on the triacylglycerol. Alternatively, there may be one enzyme which is non-specific for these two ester classes. In adipose tissue there is the well-established triacylglycerol lipase, which is known to be activated by various endocrine signals via a cAMP-dependent protein kinase and an enzymic phosphorylation reaction [45]. In adipose tissue there is also a hormone-sensitive cholesterol ester hydrolase which appears to be strikingly similar to the enzymes found in the cytosol of the adrenal cortex. The adrenal cortex and adipose tissue both contain specific receptors for ACTH in their plasma membranes, and both respond by activation of triacylglycerol lipase and cholesterol ester hydrolase through a cAMP + ATP-linked phosphorylation reaction. Likewise, both enzymes can be deactivated by a phosphoprotein phosphatase enzyme found in the cytosol of these cells.

Dietary Lipid Composition and Adrenal Function

The lipid composition of adrenal cortex and adipose tissue responds to alterations in the lipid content and composition of the diet. In our studies we fed rats on a diet containing our stock diet of 25% skimmed milk powder, 5% dried yeast and 70% wholemeal flour, supplemented in selected cases with 25% by weight of corn oil, olive oil or rape-seed oil. In most of our studies, the animals were maintained on these diets for at least 6 weeks. The rape-seed oil used in the first experiments contained 60% erucate, the olive oil contained 61% oleate, and the corn oil contained 51% linoleate.

The cholesterol ester composition of the lipid droplets in the adrenals was markedly altered by these diets. As shown in table I, on the stock low-fat diet, the predominant cholesterol esters were palmitate and oleate; on the olive oil diet, the predominant cholesterol esters were again oleate and palmitate; on the corn oil diet, the predominant esters were linoleate and oleate, and on the erucate diet, the predominant cholesterol esters were oleate and erucate.

Table I. Effect of various diets on the fatty acid composition of adrenal cortex cholesterol esters

Fatty acid	Diet				
	stock diet	stock diet and olive oil	stock diet and corn oil	stock diet and rape seed oil	
				A	B
14:0	4	1	2	2	1
16:0	29	16	14	14	18
16:1	7	3	7	5	4
18:0	7	7	8	2	4
18:1	25	41	16	33	33
18:2	13	15	18	11	13
20:1	2	3		9	3
20:4	9	7	15	3	7
22:1		1		19	10
22:4	4	5	14	2	7

Effect of Dietary Lipids on Adrenal Cortical Cholesterol Ester Hydrolase

The animals fed the stock diet, the olive oil diet or the corn oil diet all responded to elevated plasma ACTH with an elevation of the adrenal cytosolic cholesterol ester hydrolase enzyme, and in general it was possible to increase the activity of the enzyme at least two-fold. Similarly, the addition of cAMP + ATP to the adrenal cytosolic preparation *in vitro* increased the activity of the cholesterol ester hydrolase, showing that the enzyme could be activated *in vivo* and *in vitro*. By contrast, the animals fed the high erucate diet, when subjected to ACTH treatment, failed to produce an increment in the adrenal cytosolic cholesterol ester hydrolase activity [46]. Similarly, when the adrenal cytosolic fraction from the animals fed a high erucate diet was subjected *in vitro* to the addition of cAMP + ATP, there was no significant increment in the activity of the cholesterol ester hydrolase [46]. These results are summarised in figure 6.

The specificity of the adrenal cytosolic cholesterol ester hydrolase for the different cholesterol esters was investigated. The enzyme appears to have a slightly higher affinity for cholesterol linoleate than for cholesterol oleate.

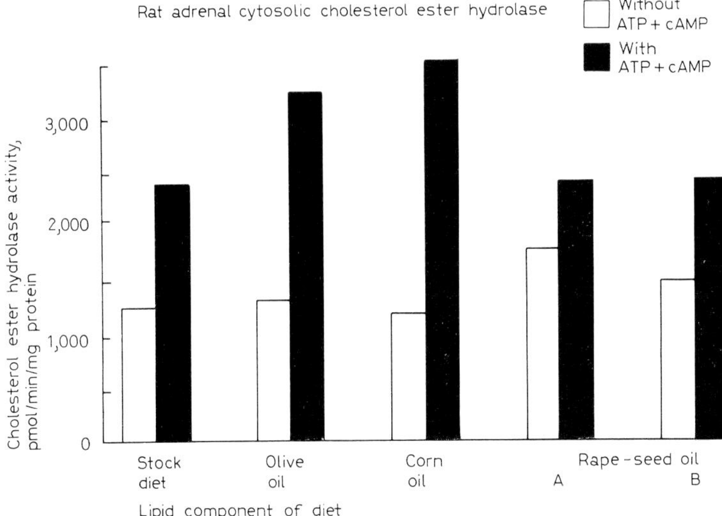

Fig. 6. Rats were fed the stock diet, a low fat diet, or this diet supplemented with olive oil, corn oil, rape-seed oil (high in erucate; A) or rape-seed oil (low in erucate; B) for a period of 6 weeks [46]. The adrenals were removed and the adrenal cytosolic cholesterol ester hydrolase measured [44] in the absence of co-factors ATP and cAMP and in the presence of ATP and cAMP. Rape-seed oil diet A had twice as much erucate as rape-seed oil B.

The enzyme has a much higher affinity for cholesterol oleate than for cholesterol stearate or for cholesterol erucate.

When a modified rape-seed oil diet was used, in which the content of erucate was about half of that used in the initial experiments [46], the accumulation of cholesterol erucate in the lipid droplets in the adrenal cortex was much less and in this case the cholesterol ester hydrolase in the cytosol of the adrenals could be activated in the normal fashion.

When animals are fed diets containing erucate, some of the fatty acid tends to accumulate in the adrenal cortex as cholesterol erucate. The cause of the accumulation of the cholesterol erucate may be an inability of the cytosolic cholesterol ester hydrolase to split this ester at a sufficient rate to prevent its accumulation. On the high erucate diets another factor was obviously operating, because there was a loss of the ability of the adrenal cytosolic cholesterol ester hydrolase to be activated by cAMP + ATP and the protein kinase.

The composition of the cholesterol esters present as lipid droplets in the adrenal cortex of the rat is influenced by the lipid composition of the diet. Rats fed a low fat diet which was marginally deficient in essential fatty acids accumulated $C_{18:1}$ and $C_{20:3}$ as the predominant fatty acids esterified to cholesterol found in the adrenal cortex. On the other hand, rats fed on an EFA-replete diet (corn oil) had mainly $C_{20:4}$ and $C_{18:1}$ as the predominant fatty acids esterified to cholesterol in the adrenal cortex. Both the EFA-deficient and the EFA-replete animals were able to respond to ACTH (or stressful situations) by switching on the hormone-sensitive cholesterol ester hydrolase in the cytosol of the adrenal cells. The inclusion of a low amount of erucic acid in the diet did not influence the hydrolase response of the animals to ACTH. However, a high percentage of erucate in the diet impaired the response of the adrenal cortex cholesterol ester hydrolase to ACTH.

Effect of Dietary Lipids on the Ability of Isolated Rat Adrenal Cortical Cells to Synthesise Corticosterone

Rats were fed on stock diet, olive oil diet, corn oil diet or rape-seed oil diets for periods of at least 6 weeks. The animals were then killed and the adrenals quickly removed, trimmed and subjected to an enzymic digestion procedure in order to isolate adrenal cortical cells. The viability of the cells was checked in the usual way and the cells were incubated in the presence and in the absence of ACTH. Corticosterone production was measured over a 60-min period. These animals responded to changes in the diet by an alteration to their adrenal cortical cholesterol ester composition (table I) but the ability to activate the adrenal cortical cytosolic cholesterol ester hydrolase was not impaired by this dietary regimen except in the case of the very high erucate diet as shown in table I. Similarly, isolated adrenal cells from rats fed a marginally deficient essential fatty acid diet, an olive oil diet, a corn oil diet or the lower erucate rape-seed oil diet all had the ability to produce normal amounts of corticosterone when incubated *in vitro*. Furthermore, these adrenal cells all responded to ACTH by producing an increment in the corticosterone output. It must be concluded therefore that alterations to the lipid composition of the diet of these rats resulted in a change in the cholesterol ester composition of the lipid droplets in the cytoplasm of the adrenal cells. However, these changes did not impair the ability of the adrenal cells to synthesise and secrete normal amounts of the main glucocorticoid and did not impair the response of these cells to ACTH.

Acknowledgements

This work was supported by a Programm Grant from the Medical Research Council.

References

1   Eisenstein, A.B.: The adrenal cortex (Little, Brown, Boston 1967).
2   McKerns, K.: Functions of the adrenal cortex, vol. 1, 2 (Appleton Century Crofts, New York 1968).
3   Symington, T.: Functional pathology of the human adrenal gland (Churchill-Livingstone, Edinburgh 1969).
4   Greep, R.O. and Astwood, E.B.: Endocrinology – adrenal gland; in Handbook of physiology, vol. 6, sect. 7 (American Physiological Society, Washington 1975).
5   Thorn, G.W.: Symposium on the Adrenal Cortex. Am. J. Med. *53:* 529 (1972).
6   Hofmann, K.; Windgender, W., and Finn, F.M.: Correlation of adrenocorticotrophic activity of ACTH analogs with degree of binding to an adrenal cortex particulate preparation. Proc. natn. Acad. Sci. USA *67:* 829 (1970).
7   Seelig, S.; Sayers, G.; Schwyzer, R., and Schiller, P.: Isolated adrenal cells: $ACTH_{11-24}$, a competitive antagonist of $ACTH_{1-39}$ and $ACTH_{1-10}$. FEBS Lett. *19:* 232 (1971).
8   Sayers, G.; Sayers, M.A.; Liang, T.Y., and Long, C.N.H.: The effect of pituitary adrenotrophic hormone on the cholesterol and ascorbic acid content of the adrenal of the rat and the guinea pig. Endocrinology *38:* 1 (1946).
9   Renold, A.E. and Cahill, G.F.: Adipose tissue; in Handbook of physiology, vol. 5 (American Physiological Society, Washington 1965).
10 Stachenko, J. and Giroud, C.J.P.: Functional zonation of the adrenal cortex site of ACTH action. Endocrinology *64:* 743 (1959).
11 Golder, M.P. and Boyns, A.R.: The location of adrenocorticotrophin receptors. Biochem. J. *129* (1972).
12 Tait, S.A.S.; Schulster, D.; Okamoto, M.; Flood, C., and Tait, J.F.: Production of steroids by *in vitro* superfusion of endocrine tissue. II. Steroid output from bisected whole, capsular and decapsulated adrenals of normal intact, hypophysectomised-nephrectomised rats as a function of time of superfusion. Endocrinology *86:* 360 (1970).
13 Brecher, P.I.; Pyun, H.Y., and Chobanian, A.V.: Studies on the angiotensin II receptor in the zona glomerulosa of the rat adrenal gland. Endocrinology *95:* 1026 (1974).
14 Krieger, D.T.; Allen, W.; Rizzo, F., and Krieger, H.P.: Characterisation of the normal temporal pattern of plasma corticosteroid levels. J. clin. Endocr. *32:* 266 (1971).
15 Berson, S.A. and Yalow, R.S.: Radioimmunoassay of ACTH in plasma. J. clin. Invest. *47:* 2725 (1968).
16 Deuel, H.J.: Lipids II (Interscience, New York 1955).
17 Richardson, M.C. and Schulster, D.: Corticosteroidogenesis in isolated adrenal cells: effect of adrenocorticotrophic hormone, adenosine 3′5′-monophosphate and β^{1-24} adrenocorticotrophic hormone diazotised to polyacrylamide. J. Endocr. *55:* 127 (1972).

18 Grahame-Smith, D.G.; Butcher, R.W.; Ney, R.L., and Sutherland, E.W.: Adenosine 3'5'-monophosphate as the intracellular mediator of the action of adrenocortico-trophic hormone on the adrenal gland. J. biol. Chem. *242:* 5535 (1967).

19 Mackie, C.; Richardson, M.D.; Schulster, D., and Robinson, C.: Kinetics and dose-response characteristics of adenosine 3'5'-monophosphate production by isolated rat adrenal cells stimulated with adrenocorticotrophic hormone. FEBS Lett. *23:* 345 (1972).

20 Robison, G.A.; Bitcher, R.W., and Sutherland, E.W.: Cyclic AMP (Academic Press, New York 1971).

21 Karaboyas, G.C. and Koritz, S.B.: Identity of the site of action of 3'5'-adenosine monophosphate and adrenocorticotropic hormone in corticosteroidogenesis in rat adrenal and beef adrenal cortex slices. Biochemistry, N.Y. *4:* 462 (1965).

22 Gill, G.N. and Garren, L.D.: Role of the receptor in the mechanism of action of adenosine 3', 5'-cyclic monophosphate. Proc. natn. Acad. Sci., USA *68:* 786 (1971).

23 Garren, L.D.; Gill, G.N.; Masui, H., and Walton, G.M.: On the mechanism of action of ACTH. Recent Prog. Horm. Res. *27:* 433 (1971).

24 Krebs, E.G. and Beavo, J.A.: Phosphorylation-dephosphorylation of enzymes. A. Rev. Biochem. *48:* 923 (1979).

25 Huijing, F. and Lee, E.Y.C.: Protein phosphorylation in control mechanisms (Academic Press, New York 1973).

26 Nimmo, H.C. and Cohen, P.: Hormonal control of protein phosphorylation. Adv. cyclic Nucleotides Res. *8:* 145 (1977).

27 Mitani, F. and Horie, S.: Studies on P-450. J. Biochem., Tokyo *65:* 269 (1969).

28 Ramseyer, J. and Harding, B.W.: Solubilisation and properties of bovine adrenal cortical cytochrome P-450 which cleaves the cholesterol side-chain. Biochim. biophys. Acta *315:* 306 (1973).

29 Wang, H.P. and Kimura, T.: Purification and characterisation of adrenal cortex mitcchondrial cytochrome P-450 specific for cholesterol side-chain cleavage activity. J. biol. Chem. *251:* 6068 (1976).

30 Hume, R. and Boyd, G S.: Biochem. Soc. Trans. *6:* 893 (1978).

31 Simpson, E.R. and Boyd, G.S.: The cholesterol side-chain system of bovine adrenal cortex. Eur. J. Biochem. *2:* 275 (1967).

32 Wilson, L.D. and Harding, B.W.: Studies on adrenal cortical cytochrome P-450. III. Effects of carbon monoxide and light on steroid 11β hydroxylation. Biochemistry, N.Y. *9:* 1615 (1970).

33 Burstein, S.; Zamoscianyk, H.; Kimball, H.L.; Chaudhuri, N.K., and Gut, M.: Transformation of labelled cholesterol, 20α-hydroxycholesterol, (22R)-22-hydroxy-cholesterol and (20R, 22R) 20,22 dihydroxycholesterol, by adrenal acetone-dried preparations from guinea pigs, cattle and men. I. Establishment of radiochemical purity of products. Steroids *15:* 13 (1970).

34 Boyd, G.S. and Trzeciak, W.H.: Cholesterol metabolism in the adrenal cortex. Studies on the mode of action of ACTH. Ann. N.Y. Acad. Sci. *212:* 361 (1973).

35 Bell, J.J.; Cheng, S.C., and Harding, B.W.: Control of substrate flux and adrenal cytochrome P-450. Ann. N.Y. Acad. Sci. *212:* 290 (1973).

36 Mason, J.I.; Arthur, J.R., and Boyd, G.S.: Regulation of cholesterol metabolism in rat adrenal mitochondria. Mol. cell. Endocrinol. *10:* 209 (1978).

37 Trzeciak, W.H. and Boyd, G.S.: Activation of cholesterol esterase in bovine adrenal cortex. Eur. J. Biochem. *46:* 201 (1974).

38 Beckett, G.J. and Boyd, G.S.: Purification and control of bovine adrenal cortical cholesterol ester hydrolase and evidence for the activation of the enzyme by a phosphorylation. Eur. J. Biochem. *72:* 223 (1977).

39 Wallat, S. and Kunan, W.H.: *In vitro* activation of a soluble cholesterol esterase from bovine adrenals by cAMP-dependent protein kinase. Hoppe-Seyler's Z physiol. Chem. *357:* 949 (1976).

40 Naghshineh, S.; Treadwell, C.R.; Gallo, L.L., and Vahouny, G.V.: Protein kinase-mediated phosphorylation of a purified sterol ester hydrolase from bovine adrenal cortex. J. Lipid Res. *19:* 561 (1978).

41 Ferguson, J.J.: Puromycin and adrenal responsiveness to adrenocorticotropic hormone. Biochim. biophys. Acta *57:* 616 (1962).

42 Ferguson, J.J.: Protein synthesis and adrenocorticotropin response. J. biol. Chem. *238:* 2754 (1963).

43 Schulster, D.; Richardson, M.C., and Palfreyman, J.W.: The role of protein synthesis in adrenocorticotropin action: effects of cycloheximide and puromycin on the steroidogenic response of isolated adrenocortical cells. Mol. cell. Endocrinol *2:* 17 (1974).

44 Gorban, A.M.S. and Boyd, G.S.: ACTH activation of cytosol triglyceride hydrolase in the adrenal of the rat. FEBS Lett. *79:* 54 (1977).

45 Pittman, R.C.; Khoo, J.C., and Steinberg, D.: Cholesterol esterase in rat adipose tissue and its activation by cyclic adenosine 3′:5′-monophosphate-dependent protein kinase. J. biol. Chem. *250:* 4505 (1975).

46 Beckett, G.J. and Boyd, G.S.: The effect of dietary rapeseed oil on cholesterol-ester metabolism and cholesterol-ester-hydrolase activity in the rat adrenal. Eur. J. Biochem. *53:* 335 (1975).

Prof. G.S. Boyd, Department of Biochemistry, University of Edinburgh,
Teviot Place, Edinburgh, EH8 9AG (Scotland)

Nutr. Metab. *24* (Suppl. 1): 90–104 (1980)

Dietary Fats and Platelet Function in French and Scottish Farmers[1]

S. Renaud, E. Dumont, F. Godsey, R. Morazain, C. Thevenon and E. Ortchanian

Inserm, Unit 63, Lyon-Bron, and Department of Nutrition, University of Montreal, Montreal

Abstract. Although the intake of saturated facts still appears to be the environmental factor most closely assosiated with coronary heart disease (CHD), this does not necessarily mean that CHD is caused essentially or solely by blood lipids, as suggested by several investigators. It seems that blood platelets rather than (or at least in addition to) blood lipids might be the intermediate link between certain environmental factors (saturated fats, hard water) and CHD, through an effect on both thrombosis and atherosclerosis. Our recent studies in French and Scottish farmers, have shown that blood platelet function is more drastically affected by saturated fats than blood lipids. In those studies, platelet function was the only blood parameter correlated on an individual basis with the intake of saturated fat and inversely related to calcium intake. Calcium is probably the cation responsible for the protective effect of hard water against CHD in various countries. The results obtained also indicate that platelet function can be improved by increasing the intake of polyunsaturated fats at the expense of saturated fats. Finally, only platelet function was different from one region of France to another and from one region of Scotland to another; this difference could be related to the reported incidence of CHD in these various regions.

Introduction

Although it is well established that severe hyperlipoproteinemia is associated with premature atherosclerosis and coronary heart disease (CHD), the relationship between serum cholesterol below 250 mg/dl and CHD is far from unanimously accepted [1].

[1] Supported in part by grants from the Inserm (Contrat CNAM), Cetiom and Fondation pour la Recherche Médicale (France) and by the Tobacco Council of Canada.

In recent years, it has been shown that the incidence of CHD was three times higher in Edinburgh than in Stockholm although the level of total cholesterol or even of HDL cholesterol was similar in the two cities [2]. In the Framingham [3] and Tromsø [4] prospective studies, total cholesterol was of no predictive value for CHD. By contrast, it seems that saturated fat consumption could be the main predisposing factor for CHD [5]. In addition to the multiple animal studies which clearly show the atherogenic and thrombogenic properties of saturated fats [7–9, the observation on the Japanese is particularly striking from this point of view. In Japan the incidence of CHD [10] is extremely low despite high incidences of hypertension, diabetes and smoking as many cigarettes (2,250 cigarettes/year) as North Americans. But in the Japanese living in the USA where they adopt the American diet [lipid calories passing from 15 to 38%; 11], the incidence of CHD becomes comparable to that observed in the USA. Consequently, it appears that saturated dietary fats are the *essential predisposing factor for CHD.*

Since dietary fats, but not serum lipids, seem to play such an essential role in CHD, the question has *arisen whether another factor rather than serum lipids* might be the intermediate link between saturated fats and CHD.

In recent years, results have accumulated indicating that blood platelets rather than blood lipids might be the blood mediator involved. In fact, blood platelets are not only involved in the early events of arterial thrombosis (occlusive and mural) which significantly contributes to CHD [12, 13], but also in the early atherosclerotic lesions. Platelets are frequently observed to migrate into the aortic wall in the early alterations observed in rabbits fed saturated fat (fig. 1); they contain a potent mitogen which stimulates the proliferation of smooth muscle cells [14]. More recent studies have indicated that this mitogen enhanced LDL receptor activity [15] and cholesterol esterification [16] of human fibroblasts, suggesting a further prominent role in the pathogenesis of atherosclerosis.

In animals, several investigators have shown that feeding saturated fats predisposed to various types of thrombosis [8, 9]. In our experimental model [8, 17, 18] as well as in that of *Hornstra* [9], it has been shown that the saturated fats appear to act mostly at the level of blood platelets through two mechanisms: (1) by accelerating coagulation through increased activity of the platelet phospholipids (PF$_3$); (2) by increasing the response of platelets to thrombin-induced aggregation. Our studies on coronary patients [19] have also indicated that their platelet phospholipids were markedly more active than in the controls; that the response of platelets to thrombin was also increased. However, there are several difficulties in studying platelet function

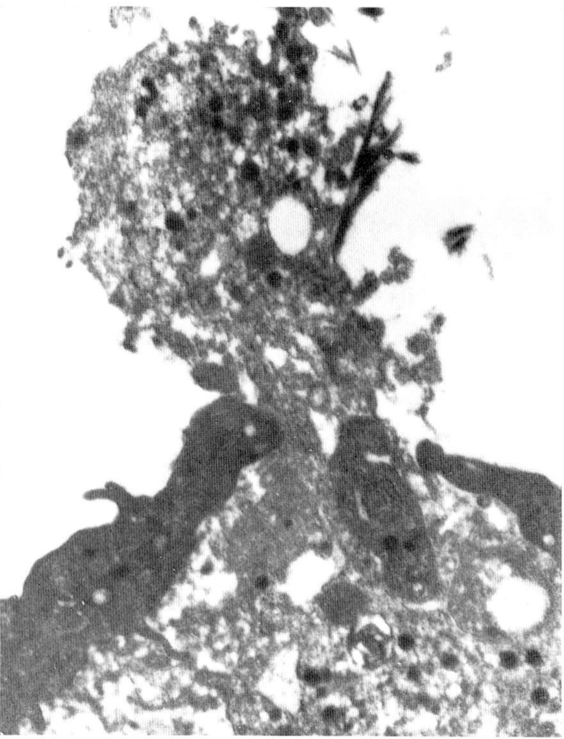

Fig. 1. One platelet entering between two endothelial cells, into the aortic wall of a butter-fed rabbit. Some more platelet material can be recognized deeper in the intima. Electron micrograph. × 6,500.

in coronary patients as compared to controls: the patients have usually extensive medications, frequently associated with modifications in the dietary habits. This usually results in drastic changes in platelet function tests; it is almost impossible to be certain that the controls in the same surrounding, are free from atherosclerotic lesions. Instead of pursuing this type of studies, we have recently compared healthy subjects from regions or countries differing essentially in the incidence of CHD.

Studies in Farmers from Two Regions of France

Farmers were selected for the following reasons: at least in Europe, male farmers (owners of a farm, 40–45 years of age) have a similar way of life

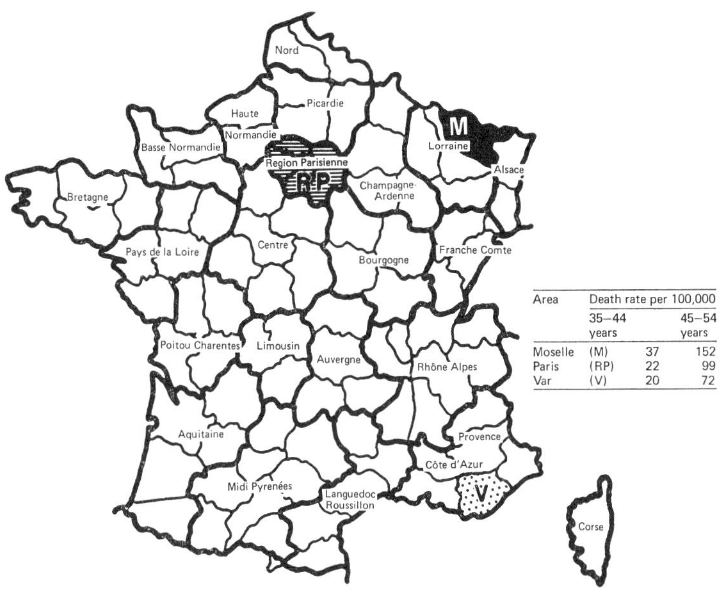

Area		Death rate per 100,000	
		35−44 years	45−54 years
Moselle	(M)	37	152
Paris	(RP)	22	99
Var	(V)	20	72

Fig. 2. Compared incidence of ischemic heart disease from 1971 to 1973 in men of three French regions. According to French Official Statistics [26].

(physical activity) and consequently are highly comparable from one region or country to another. Only the dietary habits appear to be different. In a given area, the dietary habits are quite homogeneous as are most of the other parameters examined. The farmers still keep most of the traditional habits of the region. Since all of them take their meals at home, it is easier to evaluate precisely the food pattern which does not change much from one day to another.

In France, two regions (Moselle and Var) were selected (fig. 2) because of the known differences: (1) in the mortality rate from CHD as indicated the official statistics and (2) in the dietary habits.

All the studies were performed by the same team (nurse-technician, technician, dietician); only the physical examination was performed by a local physician. Most of the work was done in a mobile laboratory, that is with the same instruments throughout the different studies. Samples of serum, plasma ,platelets and of the diet composite, were frozen for later determinations, either in the 'Laboratoire Municipal de Bordeaux' for the food chemical analysis, or in the Inserm, Unit 63, for the other determinations.

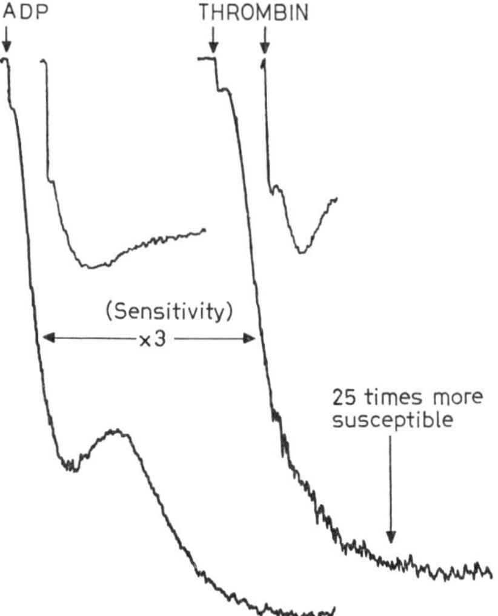

Fig. 3. Extreme differences in platelet aggregation between Moselle and Var male subject, as evaluated in a recording aggrego-coagulometer. The sensitivity of the instrument, for the highly responsive subjects, is three times less than for the other subjects. It can be calculated that the platelets from the Moselle subject are 25 times more susceptible to thrombin-induced aggregation, although serum cholesterol was similar in the two subjects.

The first study performed in the summer of 1976 [20], indicated that although there was no significant difference in serum cholesterol, between the two regions, the platelet clotting activity as well as thrombin and ADP were markedly higher in the Moselle (fig. 3), in relation to a higher intake of saturated fats (16% of calories as compared to 11% in the Var.) In addition, a highly significant correlation was obtained on an individual basis, between the intake of saturated fat and most of the platelet functions examined. This suggests that platelet function is more closely related to the intake of saturated fats [the environmental factor the most closely associated with CHD; 21] than serum cholesterol.

This first study was repeated, this time in winter, in 1977–1978, in larger samples of farmers from the same two regions. As in the summer of 1976, a highly significant difference between the two regions was observed in the clotting time of platelet-rich plasma (due to the clotting activity of platelets), and the response of platelets mostly to thrombin aggregation (fig. 4). By contrast, serum total cholesterol, triglycerides and HDL cholesterol, were

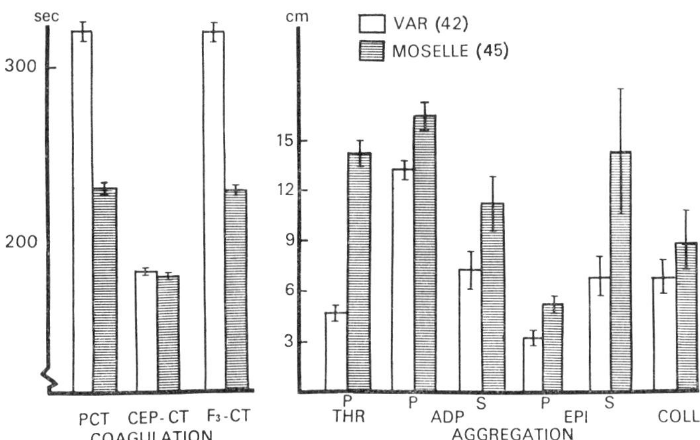

Fig. 4. Compared hematologic results between farmers (males, 40–45 years; means ± SE). From two French regions, the Var and Moselle. The study was performed in the winter of 1977–1978, Coagulation: PCT = recalcification clotting time of platelet-rich plasma (evaluates the whole blood-clotting activity); CEP-CT = cephalin clotting time of platelet-poor plasma (evaluates the clotting activity of the plasmatic clotting factors); F₃-CT = realcification clotting time of the platelets, washed and resuspended in a standard platelet-poor plasma (determines platelet factor 3 clotting activity). Aggregation in PRP: THR = thrombin-induced; ADP = induced aggregation to ADP (P, primary; S, secondary); EPI = induced aggregation to epinephrine (P, primary; S, secondary); COLL = collagen-induced.

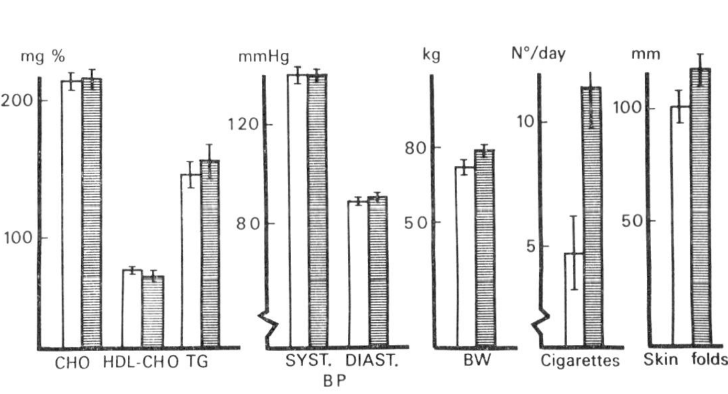

Fig. 5. Comparison between the Var and Moselle concerning various parameters investigated in the farmers studied in the winter of 1977–1978. Means ± SE. There are no significant differences for any of the parameters reported in the figure. CHO = Total cholesterol; HDL-CHO = HDL cholesterol; TG = triglycerides; Syst. = systolic; Diast. = diastolic.

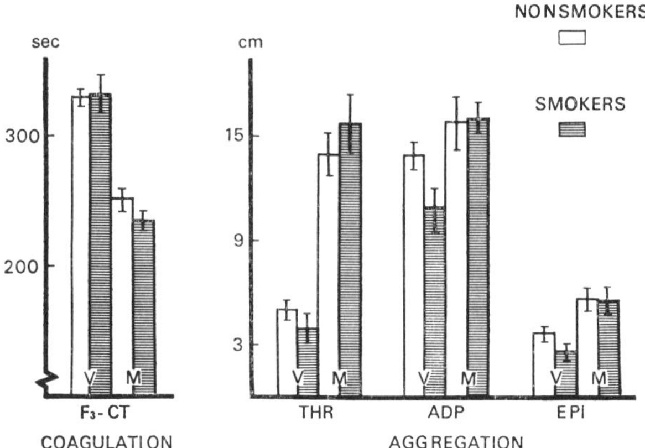

Fig. 6. Comparison in the hematologic results between nonsmokers (26 from the Var, V, and 20 from the Moselle, M, and smokers (16 from the Var and 25 from the Moselle). No significant difference could be observed between smokers and nonsmokers. The tests were performed on fasted subjects not having smoked since the day before. Means ± SE. Abbreviations as in figure 4.

identical in the two regions as was the blood pressures, body weights and skin folds (fig. 5). The number of cigarettes smoked in the Moselle was higher but, as shown in figure 6, no difierence could be observed in the hematologic parameters studied between nonsmokers and smokers, fasted and not having smoked since the day before. Our more recent studies in Scotland have indicated that the effect of smoking on platelet function is an acute one ,which can be observed within 10 min after smoking a cigarette and lasts less than 1 h (to be reported elsewhere).

As concerns the dietary habits, chemical analysis of the diet composite (fig. 7) as in the summer study of 1976, indicated that in the Moselle, there was a higher intake of lipids, mostly of saturated fats. The correlations between the blood and diet parameters appear to be of special interest (table I). Serum cholesterol, as in other studies [22], is not significantly correlated on an individual basis with the intake of saturated fats. By contrast, it can be noted that the two platelet function tests shown in animals to be most closely associated with dietary saturated fats [F_3-CT, thrombin aggregation; 17, 18], are significantly correlated with the saturated fat intake. In addition, there was an inverse significant correlation between calcium intake and platelet clotting activity (F_3-CT). By combining these two parameters (saturated fats and calcium), the correlation coefficients became ectremely significant with

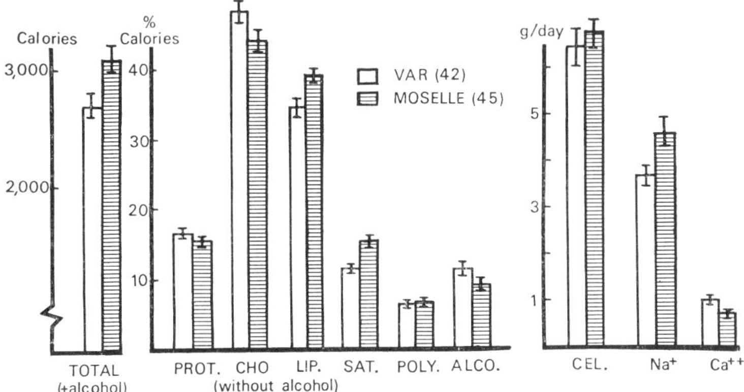

Fig. 7. Chemical analysis of the food intake in the farmers from the Var and Moselle studied in the winter of 1977–1978. Means±SE. There is a highly significant (p<0.01) difference in the intake of saturated fatty acids ,between the two regions. PROT = proteins; CHO = carbohydrate; LIP = total lipids; SAT. = saturated fatty acids; polyunsat. = polyunsaturated fatty acids; CEL. = cellulose.

Table I. Correlations (r) between blood and diet parameters on the 86 subjects from the Moselle and Var studied in the winter of 1977–1978

	Induced aggregation					Cholesterol	
	F_3–CT	THR	ADP	EPI	COLL	total	HDL
Sat., %cal	0.23	0.21	−0.03	−0.07	−0.08	−0.11	−0.13
Ca^{++}, g/day	−0.22	−0.19	−0.07	0.08	0.04	−0.17	0.04
Sat–10 × Ca^{++}	0.34	0.29	0.03	0.02	−0.09	0.07	−0.13

r = 0.33, p< 0.001; r = 0.26, p< 0.01; r = 0.20, p< 0.05.

Sat. = Saturated fats; F_3–CT = clotting activity of platelets (PF$_3$); THR = thrombin; ADP = adenosine diphosphate; EPI = epinephrine; COLL = collagen.

both F_3-CT and thrombin aggregation, calcium having a negative sign. This indicates that calcium inhibits the effect of saturated fat on at least two platelet functions (F_3-CT and thrombin). This result, which has been confirmed in recent studies in French and British farmers (to be published elsewhere), is concordant with the known beneficial effect of hard water and consequent-

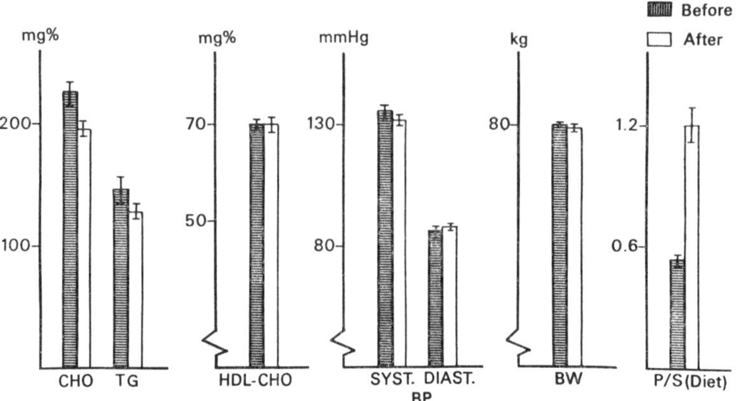

Fig. 8. Comparison between various parameters before and 1 year after diet modification in 53 Moselle farmers. Means ± SE. Abbreviations as in figure 5.

ly of calcium, on CHD [23]. It also reinforces our hypothesis on the key role of platelets in CHD.

Effects of Long-Term Diet Modification in Moselle Farmers

Not only do the Moselle farmers eat differently as compared to the Var farmers, but they are also of different ethnic origin. In order to verify that the difference in platelet function between the two regions was really due to the difference in the dietary habits, a group of 50 families were persuaded to change some of their dietary habits in such a way that they became comparable to those of the Var farmers, at least as far the intake of saturated and polyunsaturated fats is concerned. Another group of 50 Moselle families did not change their diets and served as controls. The P/S ratio passed from 0.5 to 1.25, as shown in figure 7. 1 year after the diet modification, there was no change in body weight, systolic or diastolic blood pressure and HDL cholesterol. Total cholesterol and triglycerides were decreased by approximately 10%.

The resulting changes in the fatty acid composition of plasma total lipids and platelet total phospholipids are shown in figure 9. In both plasma and platelets, the most conspicuous changes were an increase in 18:2, a slight decrease in 20:4 at the expense mostly of 16:0, 18:1 and 20:3ω9 (not shown

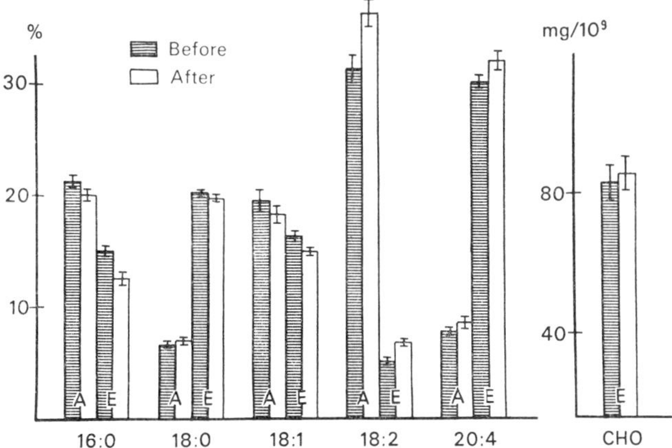

Fig. 9. Changes in plasma (A) total lipids and in platelet (E) total phospholipids fatty acid composition 1 year after diet modification. CHO = Platelet cholesterol.

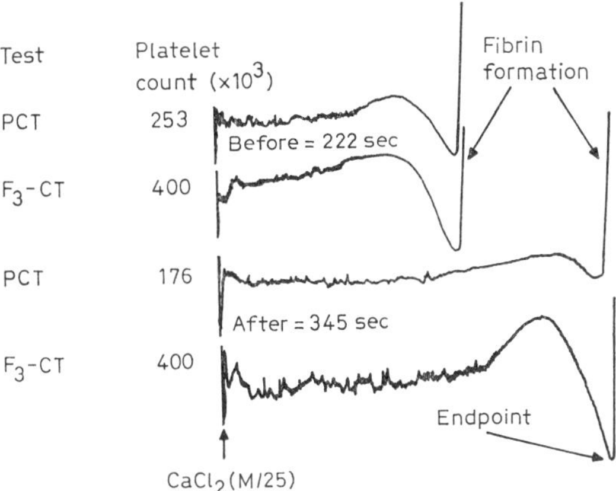

Fig. 10. PCT and F₃–CT recordings before and 1 year after diet modification. In spite of a variable platelet count in the PRP (between 253,000 and 176,000/mm³), the clotting time is identical to that obtained with platelets, washed and resuspended at a standardized count (400,000/mm³) in a platelet-poor standard plasma.

in figure 9). This result very clearly indicates that the farmers had really changed their diets, as suggested.

The hematologic results are illustrated in figure 10. Figure 11 shows the results of diet modification at 1 and 2 years. It can be seen that the clotting

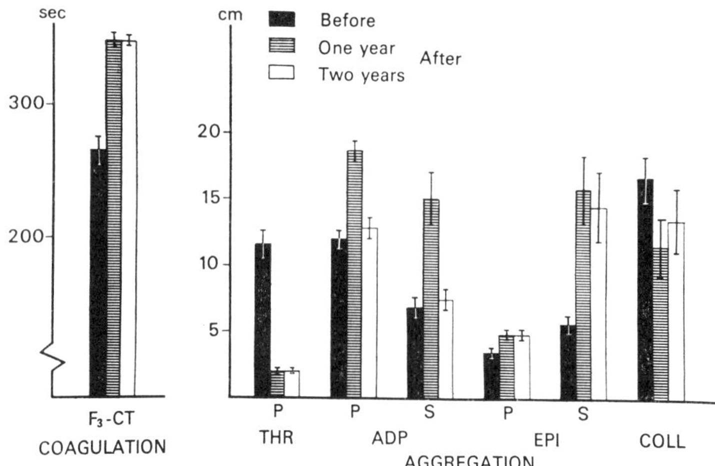

Fig. 11. Hematologic effects in 50 Moselle male farmers, 1 and 2 years after diet modification. (Means ± SE). Abbreviations as in figure 4.

activity of platelets is markedly prolonged, and the aggregation to thrombin reduced similarly after 1 and 2 years. The values obtained are now comparable or even better, than the values noted in Var farmers. There were few changes in the response to collagen. 1 year after diet modification, the response to ADP and epinephrine was markedly increased. After 2 years, ADP aggregation was back to normal, but not the susceptibility to epinephrine which was still higher than before the diet change. By contrast, the values in the control group (not reported here in details) had not changed at all after 1 year.

The present results indicate that platelet function in man depends mostly on the dietary fat intake and can be normalized within 1 year as far platelet-clotting activity and response to thrombin are concerned by decreasing the intake of saturated fats and insreasing that of polyunsaturated fats. These two factors appear to be those that are the most closely associated with the thrombotic tendency both in animals and man [17–20].

By contrast, diet modification seems to be accompanied by a higher response of platelets to ADP and epinephrine which might need several years to be decreased at the level observed in populations consuming lower amounts of saturated fat as in the Var.

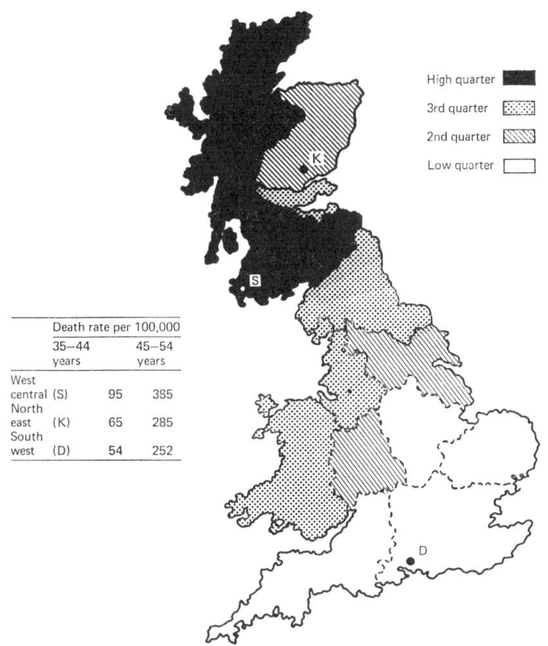

Death rate per 100,000		
	35–44 years	45–54 years
West central (S)	95	385
North east (K)	65	285
South west (D)	54	252

Fig. 12. Compared incidence of ischemic heart disease in men in Great Britain from 1969 to 1973. The 1978 investigation in Scottish farmers was performed in Kirriemuir (K) and in Stranraer (S). From *Fulton et al.* [24].

Platelet Function in Farmers from Eastern and Western Scotland

In order to verify that the results obtained in French farmers were not specific for this country, two groups of Scottish farmers were investigated in 1978.

Recent statistical analyses carried out in Scotland [24] indicated that solely the population of Western Scotland presented an extremely high incidence of CHD, while the population of the Eastern part was comparable from this point of view to England (fig. 12). Consequently, two samples of Scottish farmers were selected, one in Eastern (Kirriemuir) and the other in Western Scotland (Stranraer).

While there was no difference in any of the lipemic fractions (cholesterol, HDL cholesterol, triglycerides) studied, all the platelet function tests (fig. 13) showed a markedly higher platelet activity in Western than in Eastern Scot-

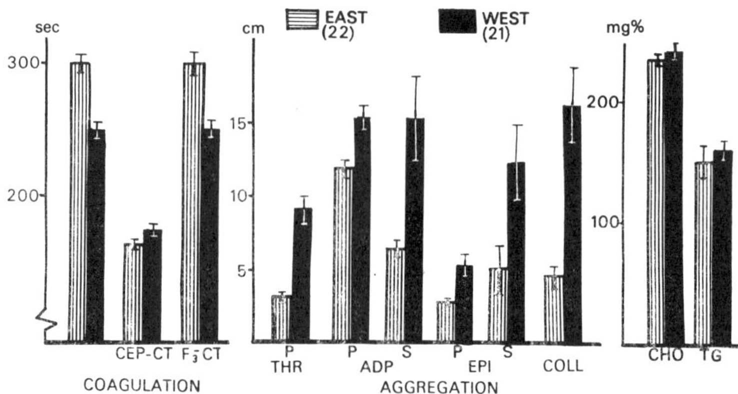

Fig. 13. Compared hematologic results in farmers from Eastern and Western Scotland (1978 study; means ± SE). Abbreviations as in figure 4.

land. The results were somewhat similar to those observed between the Moselle and Var in France.

Of interest concerning the dietary habits between Western and Eastern Scotland is that the dietary habits are absolutely identical except for the visible fats utilized on bread or for cooking. While in Western Scotland, exclusively, butter and cream are utilized (65 g/day as a mean), in the East, dairy products have been replaced, approximately 10 years ago, by vegetable oils and soft margarines (45 g/day as a mean). Analysis of 13 different brands of margarine utilized in Kirriemuir has shown a mean content in polyunsaturated fatty acids of 25%. Both the dietary and hematologic findings noted in 1978, have been confirmed by more extensive studies performed in 1979 in the same areas (to be published elsewhere).

These modifications in the dietary habits appear to be similar to those which have been studied over a period of 10 years, in the Flemish-speaking populations of Belgium [25]. In this population, the incidence of CHD seems to have considerably decreased during this period in comparison to the Walloon population of Belgium.

Consequently, if 10 years ago, the dietary habits and the incidence of CHD were similar in Western and Eastern Scotland, as is the case in Belgium, this would represent examples of CHD prevention by simple modifications of dietary habits, which might be mostly related to changes in platelet functions.

References

1   Reiser, R.: Oversimplification of diet: coronary heart disease relationship and exaggerated diets recommendations. Am. J. clin. Nutr. *31*: 865–875 (1978).

2   Oliver, M.F.; Nimmo, I.A.; Cooke, M.; Carlson, L.A. and Olsson, A.G.: Ischaemic heart disease and associated risk factors in 40 years old men in Edinburgh and Stockholm. Eur. J. clin. Invest. *5:* 507–514 (1975).

3   Castelli, W.P.; Doyle, J.T.; Gordon, T.; Hames, C.G.; Hjortland, M.C.; Hubley, S.B.; Kakan, A. and Zukel, W.J.: HDL cholesterol and other lipids in coronary heart disease. The cooperative lipoprotein phenotyping study. Circulation *55:* 757–772 (1977).

4   Miller, N.E.; Thelle, D.S.; Forde, O.H. and Mjos, O.D.: The Tromsø heart study. High density lopoprotein and coronary heart disease: a prospective case control study. Lancet *i:* 965–967 (1977).

5   Stamler, J.: Epidemiology of coronary heart disease. Med. Clins N. Am. *57:* 5–46 (1973).

6   Turpeinen, O.: Effect of cholesterol-lowering diet on mortality from coronary heart disease and other causes. Circulation *59:* 1–7 (1979).

7   Renaud, S.: Dietary fats and atherosclerosis in rat and rabbit. Adv. Cardiol. *13:* 169–182 (1974).

8   Renaud, S. and Gautheron, P.: Dietary fats and experimental (cardiac and venous) thrombosis. Haemostasis *2:* 53–72 (1973).

9   Hornstra, G.: Experimental studies. Dietary fats and arterial thrombosis; in Renaud and Nordöy, Dietary fats and thrombosis, pp. 21–52 (1973).

10 Kato, H.; Tillotson, J.; Nichaman, M.Z.; Rhoads, G. and Hamilton, H.: Epidemiologic studies of coronary heart disease and stroke in Japanese men living in Japan, Hawai and California. Am. J. Epidem. *97:* 372–385 (1973).

11 Marmost, M.G. and Syme, L.S.: Acculturation and coronary heart disease in Japanese-Americans. Am. J. Epidem. *104:* 225–247 (1976).

12 Duguid, J.B.: Thrombosis as a factor in the pathogenesis of aortic atherosclerosis. J. Path. Bact. *60:* 57–61 (1948).

13 Haust, M.D.; More, R.H., and Movat, H.Z.: The mechanism of fibrosis in arteriosclerosis. Am. J. Path. *35:* 265–273 (1959).

14 Ross, R.; Glomset, J.; Kariya, B., and Harker, L.: A platelet-dependent serum factor that stimulates the proliferation of arterial smooth muscle cells *in vitro*. Proc. natn. Acad. Sci. USA *71:* 1207–1210 (1974).

15 Chait, A.; Bierman, E.L.; Albers, J.J., and Ross, R.: Stimulation of the low density lipoprotein (LDL) receptor by platelet factor (PF). Part II. Circulation *4:* 218 (1979).

16 Witte, L.D. and Cornicelli, J.A.: Effect of platelet derived growth factor on cellular lipid metabolism. Part.II. Circulation *4:* 218 (1979).

17 Renaud, S.: Role of platelet factor 3 activity in hypercoagulable states. Thromb. Diath. haemorrh. *30:* suppl. 56, pp. 11–20 (1973).

18 Renaud, S. and Godu, J.: Induction of large thrombi in hyperlipemic rats by epinephrine and endotoxin. Lab. Invest. *21:* 512–518 (1969).

19 Renaud, S.; Gautheron, P.; Arbogast, R., and Dumont, E.: Platelet factor 3 activity and platelet aggregation in patients submitted to coronarography. Scand. J. Haematol. *12:* 85–92 (1974).

20 Renaud, S.; Dumont, E.; Godsey, F.; Supplisson, A., and Thevenon, C.: Platelet functions in relation to dietary fats in farmers from two regions of France. Thromb. Haemostasis *40:* 518–531 (1978).
21 Keys, A.: Coronary heart disease in seven countries. Circulation *41:* suppl. 1 (1970).
22 Nichols, A.B.; Ravenscroft, C.; Lamphiear, D.E., and Ostrander, L.D., Jr. :Daily nutritional intake and serum lipid levels. The Tecumseh study. Am. J. clin. Nutr. *29:* 1384–1392 (1976).
23 Masironi, R.; Pisa, Z., and Clayton, D.: Myocardial infarction registry network. Bull. Wld Hlth Org. *57:* 291–299 (1979).
24 Fulton, M.; Adams, W.; Lutz, W., and Oliver, M.F.: Regional variations in mortality from ischaemic heart and cerebrovascular disease in Britain. Br. Heart J. *40:* 563–568 (1978).
25 Joosens, J.V.; Brems-Heyns, E.; Claes, J.H.; Graffar, M.; Kornitzer, M.; Pannier, R.; Van Houte, O.; Vuylsteek, K.; Carlier, J.; De Backer, G.; Kesteloot, H.; Lequime, J.; Raes, A.; Vastesaeger, M., and Verdonk, G.: The pattern of food and mortality in Belgium. Lancet *i:* 1069–1072 (1977).
26 Statistique des causes médicales de décès (Inserm, Paris 1971–1973).

S. Renaud, Inserm, Unité 63, 22, avenue du Doyen-Lépine, F-69500 Bron (France)

Nutr. Metab. *24* (Suppl. 1): 105–118 (1980)

Favourable Influences of Linoleic Acid on the Progression of Diabetic Micro- and Macroangiopathy

A.J. Houtsmuller, J. van Hal-Ferwerda, K.J. Zahn and H.E. Henkes

Erasmus University, Department of Internal Medicine, Eye Hospital, Rotterdam

Key Words. Diabetic microangiopathy · Serum linoleic acids · Diet effects · Cholesterol · Triglycerides · Heart infarction

Abstract. A long-term study of two diets on the progression of diabetic retinopathy was performed on 102 patients. Diet 1 was composed of carbohydrates 50 cal%, saturated fats 35 cal%, proteins 15 cal%; diet 2, of carbohydrates 45 cal%, fats 40 cal% (are third linoleic acid) and proteins 15 cal%. The linoleic acid content of diet 2 was 4 times that of diet 1. The diabetics were matched 2 by 2 according to 10 criteria and set at random on one of the two diets. During the observation period of 5 years, 3 patients of diet 1 died of a heart infarction.

Of the remaining 96 patients (51 males and 44 females); of diet group 1 (linoleic acid poor), 62% of the males and 55% of the females showed retinopathy. Of diet group 2 (linoleic acid rich), 27% of the males and 32% of the females showed retinopathy (p<0.001). A strong inverse correlation was found between progression of retinopathy and serum cholesteryl linoleate.

Of diet group 1, 11 males and 7 females showed signs of cardiac ischaemia, of diet group 2, 3 males and 2 females (p<0.025) did so. Large differences were seen for the GTTs and corresponding insulin levels between males and females. While the females of diet group 2 improved remarkably in respect of GTTs and reactive serum insulins, a slow deterioration of both parameters was seen for diet group 1. Regarding the males, no differences between the two diet groups were seen. The improvements due to the treatment were similar. The serum lipid responses were significant for the females, diet group 2 showed the lowest levels; however, no differences were seen for the males. In most cases of both sexes, platelet aggregation was increased, no effects of linoleic acid was seen: an explanation for the contradictory facts was not given.

More than 50 years after the discovery of insulin, the optimal diet of patients with diabetes mellitus is still a point of controversy. In general, two diets are advocated to be the best, a carbohydrate-rich diet, low in saturated fats, and a diet lower in carbohydrates and rich in polyunsaturated fatty acids.

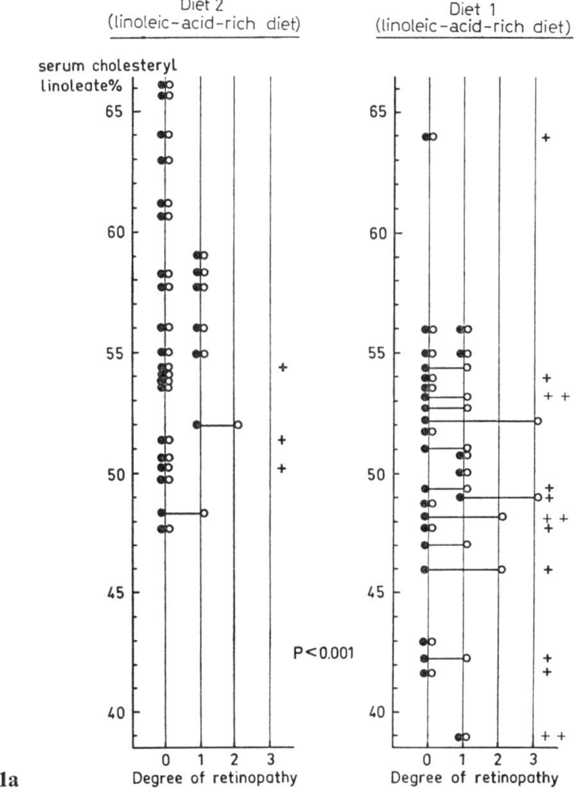

Fig. 1. Progression of retinopathy and cardiac ischaemia in diabetic patients on two different diets for 5 years. ● = Degree of retinopathy at the start of the study; O = degree of retinopathy at the end of the study; Horizontal lines = progression of the retinopathy; + = development of cardiac ischaemia; + + = development of heart infarction. **a** Males (two groups, 26 patients per group). **b** Females (two groups, 22 patients per group).

Differences in results can often be explained by regional or genetically determined heterogeneity.

After observing hypoglycaemic reactions with increasing frequency in diabetic patients set on linoleic-acid-rich diets, we showed in balance studies in insulin-independent diabetics a much lower 24-hour serum insulin level, combined with unaltered blood glucose levels after replacing a carbohydrate-rich saturated-fat diet by a linoleic-acid-rich diet [*Houtsmuller*, 1975].

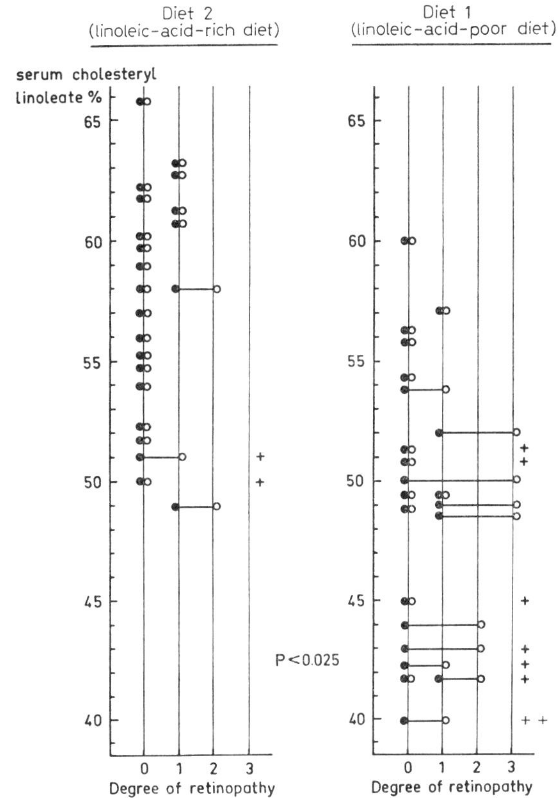

1b

This brought us to the conclusion that a linoleic-acid-rich diet is insulin sparing. So we decided to study the effects of linoleic acid on the progression of diabetic retinopathy in a long-term prospective study.

Subjects and Methods

Statistical analysis of the possible influences of diets on the progression of diabetic microangiopathy in 50 2 by 2 matched pairs of recently discovered patients with adult-onset diabetes on two different diets showed us that there would be statistically significant differences between the two groups in about 9 years. So we selected over a period of less than 1 year a large number of newly discovered patients with adult-onset diabetes, without or with mild diabetic retinopathy.

The patients were matched 2 by 2 following 10 criteria: age, sex, body weight, degree of retinopathy, glucose tolerance test, their corresponding serum insulins, serum lipids, heredity, smoking habits and therapy. All patients with abnormal blood pressure, liver, kidney, lung or heart functions, including ECG, were excluded. 102 patients were selected for the trial and divided at random in two diet groups; either a diet with carbohydrates 50 cal%, saturated fats 35 cal%, proteins 15 cal% (diet 1), or a diet with carbohydrates 45 cal%, fats 40 cal% (one third linoleic acid) and proteins 15 cal% (diet 2).

The linoleic acid content of diet 2 was 4 times that of diet 1. In all patients retinal fluorescein angiography was performed every 8 months, 50-gram GTT with corresponding insulins every 6 months, total cholesterol, triglycerides, total fats and serum cholesteryl linoleate every 3 months. All patients were examined by the same investigators and supervised by the same dietician. The fluorescein angiograms were examined blind by two independent ophthalmologists.

Based upon the fluorescein angiographic findings, the cases were classified as follows: stage 0: absence of microangiopathy; stage 1: microaneurysms only; stage 2: microaneurysms, haemorrhages, exudates and non-perfused areas; stage 3: as stage 2, but in addition pre-retinal new vessel formation. Statistical analyses were performed according to the method of *Fleiss and Everitt* [1971] and with the χ^2 test.

Results

During the first 2 years, 2 male patients died of a heart infarction; 1 female patient died after 3 years; all with mean serum cholesteryl linoleate levels lower than 40%. After 5 years, there were still 96 patients on the trial, 52 matched males and 44 matched females.

Diabetic Microangiopathy

At the beginning of the study, 25% of the patients already showed signs of mild diabetic retinopathy (12 males and 12 females; 21 and 27%); after 5 years, the numbers increased to 23 males and 19 females (41 and 43%); of these, 16 males and 12 females (62 and 55%) belonged to diet group 1 (low in linoleic acid); in diet group 2 (rich in linoleic acid), 7 males and 7 females (27 and 32%) were found (p<0.001 and p<0.025; fig. 1a, b).

A strong inverse correlation was found between progression of retinopathy and serum cholesteryl linoleate levels. Progression was clearly inhibited in patients showing serum cholesteryl linoleate levels greater than 55%.

Macroangiopathy

Although the primary aim of the study was limited to the study of microangiopathy, regular ECGs were carried out which permitted us to register macroangiopathic damage as well. The development of macroangiopathy reflected that of microangiopathy.

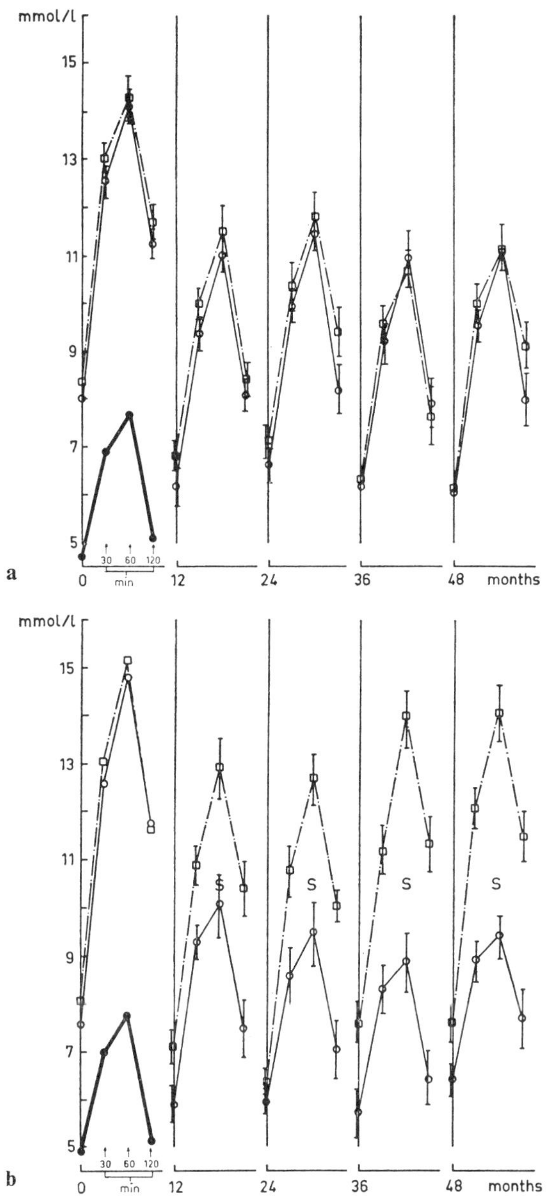

Fig. 2. 50-gram GTTs performed at 0, 30, 60 and 120 min after 12, 24, 36 and 48 months in diabetic patients. 50 normal subjects served as controls. $-\cdot-$ = Linoleic-acid-poor diet; $-$ = linoleic-acid-rich diet; ▬ = controls; S = statistically significant. **a** Males (two groups, 26 patients per group). **b** Females (two groups, 22 patients per group).

In the male group, 11 patients on diet 1 and 3 patients on diet 2 developed signs of cardiac ischaemia; 3 of these patients (all on diet 1) had heart infarction. In the female group, 7 patients on diet 1 and 2 on diet 2 were found to have cardiac ischaemia. The differences were highly significant ($p < 0.025$). 1 patient on diet 1 died of her infarction; her cholesteryl linoleate level was 39% (fig. 1a, b).

Clinical-Chemical Reactions

Glucose Tolerance Tests

Males (fig. 2a). The GTTs improved in both groups during the first months due to the treatment given. They remained at this improved level during the whole observation period. No patients needed insulin substitution.

Females (fig. 2b). The GTTs of the two groups differed markedly. While the GTTs of group 2 (linoleic acid rich) improved strongly during the observation period, those of group 1 (linoleic acid poor) deteriorated gradually after the initial improvement due to the installed therapy. Differences were highly significant ($p < 0.01$). Of these patients, 2 of group 1 and 1 of group 2 needed insulin substitution within 1 year of treatment.

Insulin Responses

The insulin responses after the GTTs were different for males and females (fig. 3a, b).

Whereas the mean GTTs of males and females at the start of the investigation corresponded well with each other, their insulin responses differed markedly, being at a normal range for the females and much lower for the males. The insulin responses for the males gradually decreased for both groups to the same degree. In sharp contrast herewith, the insulin responses for the two female groups differed markedly, decreasing gradually for group 1 (linoleic acid poor), and staying at a high level for group 2 (linoleic acid rich; $p < 0.01$).

Blood Lipids

Males (fig. 4a). The effects on the blood lipid classes were not significantly different in the two diet groups except for two periods. The initial therapeutic improvement disappeared after 1 year and was followed for 2 years by a rather large deterioration, both for total cholesterol and total lipids, however, the deterioration was worse for diet group 1 (linoleic acid poor).

Females (fig. 4b). The initial therapeutic improvement for both female groups was followed by an increasing difference between them for total

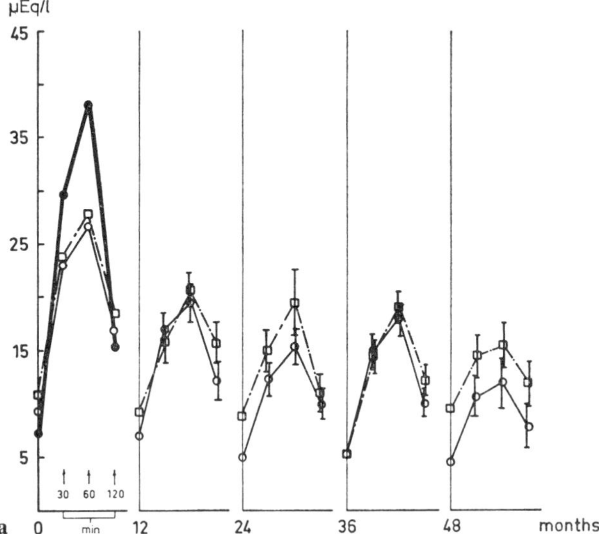
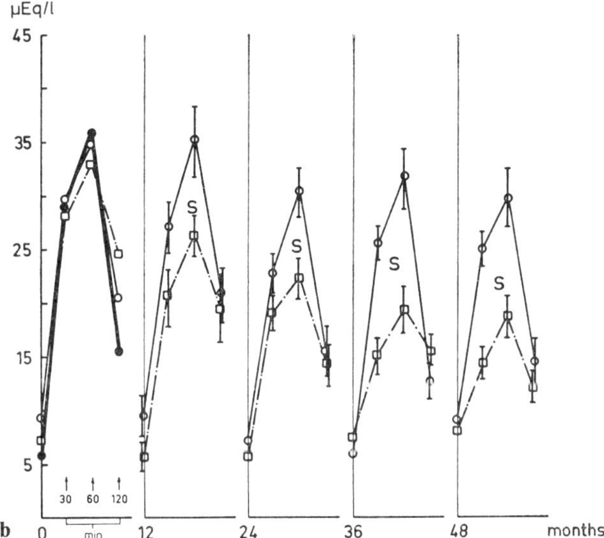

Fig. 3. Serum insulin responses after 50-gram GTTs in diabetic patients. 50 normal subjects served as controls. $-\cdot-$ = Linoleic-acid-poor diet; $-$ = linoleic-acid-rich diet; $\blacksquare$ = controls; S = statistically significant. **a** Males (two groups, 26 patients per group). **b** Females (two groups, 22 patients per group).

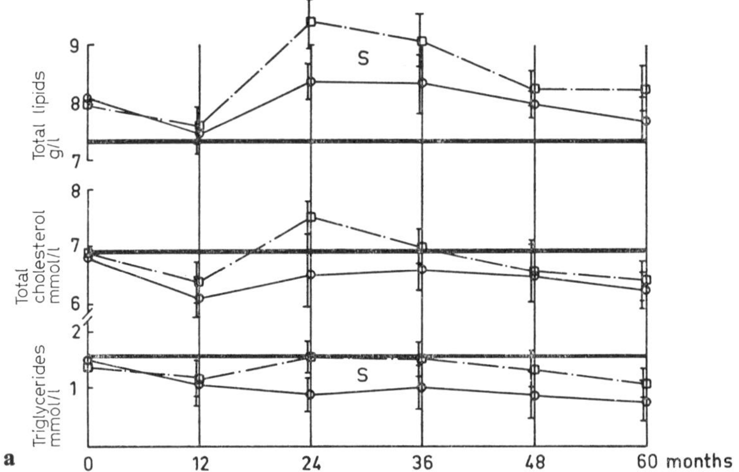

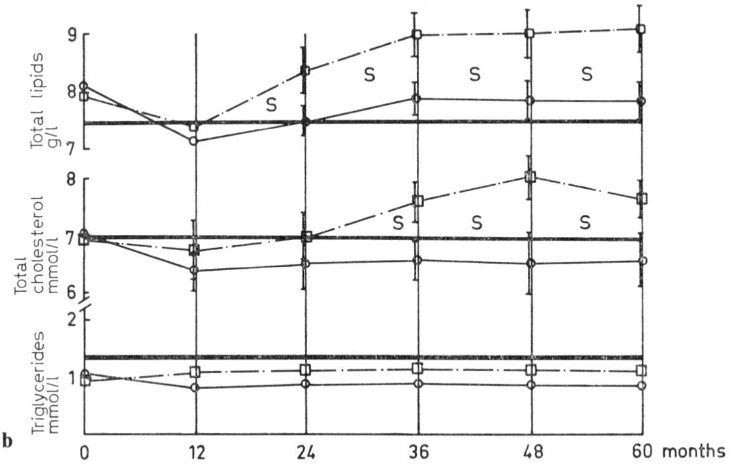

Fig. 4. Changes in total lipids, total cholesterol and triglycerides in the serum of diabetic patients on two different diets for 60 months. $-\cdot-$ = Linoleic-acid-poor diet; $-$ = linoleic-acid-rich diet; ▬ = mean of 150 normal subjects; S = statistical significance (p<0.05). **a** Males (two groups, 26 patients per group). **b** Females (two groups, 22 patients per group).

cholesterol and total lipids, which reached statistical significance after 2 years (p<0.05). Differences for triglycerides were not seen. Differences in body weight between groups were never found.

Platelet Aggregation

ADP platelet aggregation was performed once for all patients in the last year of observation. No correlation whatever could be found for platelet aggregation and linoleic acid, for progression of retinopathy, sex or age, or serum glucose, insulin or lipids.

Permanently Elevated Blood Sugar Levels and Microangiopathy

As permanently elevated blood sugar levels are considered as one of the main factors in the development of diabetic retinopathy, all patients with permanently elevated blood glucose levels were studied for expected development or progression of retinopathy. In diet group 1 (linoleic acid poor), 8 males and 5 females were found; of these, 7 males and all 5 females showed progression of retinopathy. The only male patient without retinopathy showed a mean cholesteryl-linoleate serum of 54%! In sharp contrast with this group not a single case of the 7 males and 4 females of group 2 (linoleic acid rich) showed the expected progression of retinopathy, the differences being highly significant (p<0.01).

Permanently Elevated Blood Sugar Levels and Macroangiopathy

To discover a possible correlation between permanently elevated blood sugar levels and cardiac disturbances, the same group was also considered for this relation. Of the 8 males and 5 females of diet group 1, 6 males and 3 females showed signs of cardiac ischaemia. Of the males, 2 patients were found to have heart infarction, of the females 1 patient died of her heart infarction.

Of the 7 males and 4 females of diet group 2, only 2 males showed signs of cardiac ischaemia and not a single case in the female group (p<0.01).

Discussion

Our study clearly shows a strong inhibitory effect of linoleic acid on the progression of both micro- and macroangiopathy. Several facts are of interest. Firstly, the existing retinopathy at the start of the study in 25% of recently discovered cases of maturity-onset diabetes. This suggests that part of these patients already had hyperglycaemia for some time, without overt

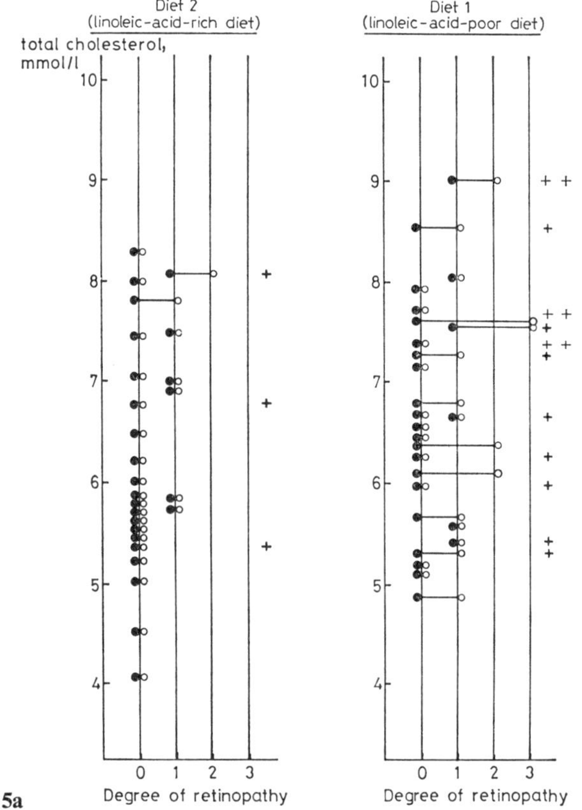

Fig. 5. Progression of retinopathy and cardiac ischaemia in diabetic patients on two different diets for 5 years. ● = Degree of retinopathy at the start of the study; ○ = degree of retinopathy at the end of the study; horizontal lines = progression of the retinopathy; + = development of cardiac ischaemia; + + = development of heart infarction. a Males (two groups, 26 patients per group). b Females (two groups, 22 patients per group.

diabetic symptoms, probably caused by the elevated renal threshold for glucose as is often seen in elderly diabetics.

The study further shows the rapid progression of retinopathy in adult-onset diabetes, it doubled within 5 years of observation. The fact that a mean serum cholesteryl linoleate concentration of over 55% strongly inhibits the development of progression of retinopathy, seems of importance even for patients with chronic elevated blood sugar levels. The same seems to hold

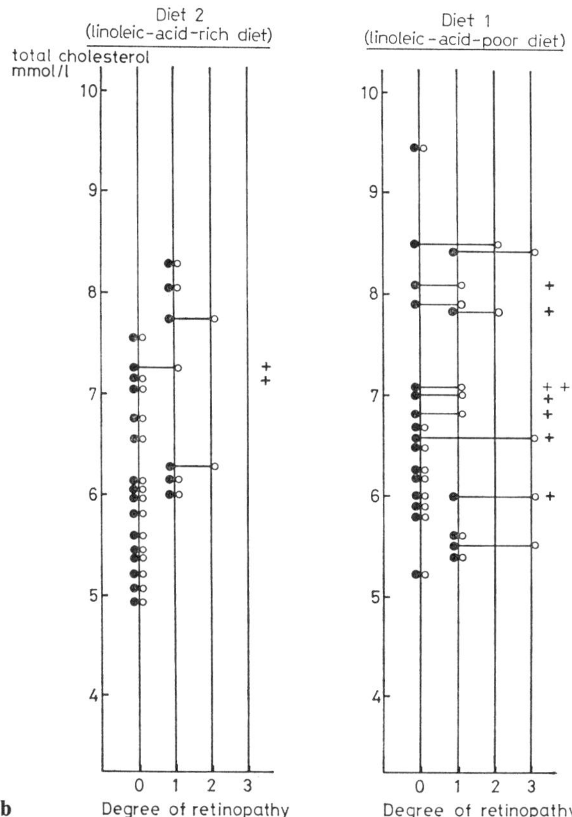

for macroangiopathy, at least for cardiac ischaemia, although the number
of patients is small.

Another point of interest is the sufficient dietary adherence, which is
reflected in the differences of the mean serum cholesteryl linoleate levels (fig.
1a, b). When considered for the mean total serum cholesterol, the difference
between the two groups is much less (fig. 5a, b).

Several authors have described changes in hormone receptors on different diets, the effects supposed to be correlated to changes in the fluidity of
membrane phospholipids [*Hülsmann,* 1977, 1978; *Awad and Zepp,* 1979].

These authors found increased sensitivity of these hormone receptors
on polyunsaturated fat diets, which was explained by increased membrane
fluidity.

Smith et al., [1974] were among the first to show that a diet, rich in saturated fats decreased the sensitivity of adipose tissue to insulin. Also an increase of the adenylate cyclase activity on a diet rich in polyunsaturated fatty acids was reported [*Orly and Schramm,* 1975].

Hülsmann [1977, 1978] showed that decrease in fatty acid desaturation in diabetes and myxoedema can be improved by unsaturated fatty acids. On the other hand, several investigators showed a decrease in prostacycline (PG J_2) production in the vessels in diabetes, causing increased thrombocyte aggregation [*Johnson et al.,* 1979; *Silberbauer et al.,* 1979]. This defect could be amended by linoleic acid.

Increased synthesis of prostaglandins by diabetic platelets was reported [*Halushka et al.,* 1977].

In balance studies in patients with maturity-onset diabetes we have shown a decrease in serum insulin, following a linoleic acid rich diet [*Houtsmuller et al.,* 1974]. This result may be explained by the results of *Smith et al.* [1974]. Also the differences found in this study in glucose tolerance tests and serum insulins in the female diabetics on diets 1 and 2 may have the same explanation, if not the behaviour of the male patients had been contradictory. For the moment it seems difficult to explain these large differences only by differences in sex hormones, while the effects on the micro- and macroangiopathy is the same for both sexes. To the best of our knowledge, such differences have not yet been reported for diabetic patients. An explanation based on the different behaviour of the platelets in the two diets cannot be excluded, because we have studied platelet aggregation only once at the end of the trial. However, the fact that we have not seen any correlation does not support this suggestion.

Up to now, we have no valid explanation for the protective effects of linoleic acid on diabetic micro- and macroangiopathy.

Further studies are in progress.

References

Awad, A.B. and Zepp, A.: Alteration of rat adipose tissue lipolylic response to norepinephrine by dietary fatty acid manipulation. Biochem. biophys. Res. Commun. *86:* 138 (1979).

Fleiss, J.L. and Everitt, B.S.: Comparing the marginal totals of square contingency tables. Br. J. Math. Statist. Psychol. *24:* 117 (1971).

Halushka, P.V.; Lurie, D., and Colwel, J.A.: Increased synthesis of prostaglandin E like material by platelets from patients with diabetes mellitus. New Engl. J. Med. *297:* 1306 (1977).

Houtsmuller, A.J.: The role of fats in the treatment of diabetes mellitus; in Vergroesen, The role of fats in human nutrition, p. 231 (Academic Press, London 1975).

Houtsmuller, A.J.; Broek-Boot, M.M.A. van den, and Peters, G.A.M.: Influence of various diets on diurnal levels of plasma glucose, insulin and lipids in obesity and insulin independent diabetics; in Schlettler and Weisel, Atherosclerosis, p. 782 (Springer, Berlin 1974).

Hülsmann, W.C.; Oerlemans, M.C., and Gulhoed-Muras, M.M.: Effects of hypothyroidism, diabetes and poly-unsaturated fatty acids on heparin releasable rat liver lipase. Biochem. biophys. Res. Commun. *79:* 784 (1977).

Hülsmann, W.C.: Dieet en insuline resistentie. Ned. Tijdschr. Geneesk. *122:* 1133 (1978).

Johnson, M.; Harrison, H.E.; Raftery, A.T., and Elder, J.B.: Prostacyclin may be reduced in diabetes in man. Lancet *i:* 325 (1979).

Orly, J. and Schramm, M.: Fatty acids as modulators of membrane functions: catecholamine activated adenylate cyclase of the turkey erythrocyte. Proc. natn. Acad. Sci. USA *72:* 3433 (1975).

Silberbauer, U.; Schernthauer, G.; Sinzinger, H.; Piza-Katzer, H., and Winter, M.: Decreased vascular prostacyclin in juvenile onset diabetes. New Engl. J. Med. *300:* 366 (1979).

Smith, U.; Kral, J., and Bjorntorp, P.: Influence of dietary fat and carbohydrate on the metabolism of adipocytes of different size in the rat. Biochim. biophys. Acta *337:* 278 (1974).

A.J. Houtsmuller, Eye Hospital, Department of Internal Medicine,
Erasmus University, Schiedamse Vest, 180, 3011 BH Rotterdam (The Netherlands)

Discussion

Prof. Avogaro: Was there a difference between the two groups as regards the glucose level, what about other diabetic parameters?

Dr. Houtsmuller: Both patient groups started with the same glucose levels, as they were matched for these parameters. 24-hour urine glucose levels could not be used as a parameter because of the high renal threshold for glucose, as is found in maturity-onset diabetes. During the study, the differences in diabetic symptoms became obvious.

Prof. Verdonk: I have clinical experience about improvement in diabetics with replacement of saturated fats by polyunsaturated fats: have you not stated that your patients with retinopathy and receiving linoleic acid needed less insulin?

Dr. Houtsmuller: We often saw hypoglycemic reactions when we administered linoleic acid to insulin-dependent diabetics; they indeed needed lesser insulin. This was observed mostly for females, as they were the main diabetic group studied. Only after finishing the study did we realize that the results for males were much less significant than for females.

Prof. Hartman: Do you have any indication that microangiopathy of other organs, e.g. the kidney, has been favorably influenced?

Dr. Houtsmuller: We have not examined the kidneys directly; however, differences in serum creatinine, serum urea or sedimentation have not been found so far.

Dr. Peeters: I would like to comment on this paper and the others. We often forget that linoleic acid is present in large amounts in phosphatidylcholine (lecithin) and it is known that in type 2 arteriosclerosis in men, the relation between the amount of lecithin and sphingomyelin is shifted to the advantage of sphingomyelin and atherosclerosis. And this is one of the reasons why you can have, for instance, a normal cholesterol level, but a different composition of the platelet phospholipids. It has been shown that especially the phospholipid composition of HDL varies tremendously in atherosclerosis where each time lecithin is decreased together with the unsaturated fatty acids, whereas sphingomyelin, which has more saturated fatty acids, is increased. Many data are rather difficult to interpret because we do not consider a parameter which might be very functional at the level of the platelets, of the arterial wall.

Dr. Houtsmuller: We had no possibility to study the phosphatidylcholine-sphingomyelin ratio in our patients.

Nutr. Metab. *24* (Suppl. 1): 119–141 (1980)

L'acide linoléique dans l'équilibre alimentaire
Recherches expérimentales

J. Lederer, A. M. Pottier-Arnould, E. Niethals, M. Dehez-Delhaye,
R. Chakroun, C. Delhaye-Pottier et M. Goovaerts

Université de Louvain, Ecole de Santé Publique, Unité de Nutrition, Bruxelles

Key Words. Linoleic acid · Triglyceridemia · Physical work · Natrium chloride

Abstract. In the male rat a diet rich in beef fat facilitates the occurrence of hyperinsulinemia after a glucose load whereas fats rich in linoleic acid produce no such effect. The combination of saturated fat and saccharose facilitates the occurrence of hypertriglyceridemia in the male rat, no such effect is produced by the combination of fats rich in linoleic acid and saccharose. Linoleic acid prevents natrium chloride from provoking hypertriglyceridemia in male and in female rats subjected to a diet enriched in saccharose and fat. Estrogen-induced hypertriglyceridemia in castrated animals is strongly inhibited if the diet is rich in linoleic acid. Physical effort can prevent saccharose combined with saturated fats from inducing hypertriglyceridemia.

But de la recherche

Dans le présent travail nous avons étudié les points suivants:

1) Influence de la qualité des graisses sur l'assimilation du glucose: *Himsworth* [1934] avait montré il y a 45 ans que l'enrichissement du régime en matière grasse provoquait chez le lapin une diminution de la tolérance au glucose; toutefois, cette expérience n'avait été menée que sur un petit nombre d'animaux sans faire de distinction de sexe ni de distinction de la qualité des graisses.

2) Influence de l'association du sucre et des graisses de différentes qualités sur l'assimilation du glucose; ceci est justifié par le fait que d'une part *Yudkin* [1957], confirmé par beaucoup d'autres, a montré que le gros

mangeur de graisse est souvent aussi un gros mangeur de sucre et que d'autre part l'excès de consommation de sucre mène au diabète.

3) Influence de la qualité des graisses sur les différents paramètres lipidiques, ceci chez les animaux des deux sexes dans le but de faire une recherche systématique sur les faits observés par *Kinsell et al.* [1956] et *Keys et al.* [1957].

4) Influence de l'association du sucre et des graisses de différentes qualités sur les différents paramètres lipidiques chez des animaux des deux sexes afin de faire une recherche systématique sur les faits observés par *Macdonald* [1972].

5) Influence des hormones sexuelles sur l'action lipidogène du sucre et des graisses de différentes qualités afin d'établir si les particularités du comportement propre à chaque sexe sont d'origine hormonale ou génétique.

6) Influence du travail physique sur l'action lipidogène de l'association graisse saturée-saccharose.

7) Influence du sel sur l'action lipidogène de certains régimes.

Matériel et méthode

Animaux
L'expérience a été conduite sur des rats Wistar de lignée pure élevée depuis plus de 30 ans dans notre laboratoire sans contamination exogène.

Régimes alimentaires
Huit régimes différents ont été expérimentés: 1) régime normal, constitué d'un aliment de base commercial, contenant 20% de protéines, 3,5% de graisse, 53,5% de sucre et d'amidon, 10% de sels minéraux et 13% d'eau; 2) régime enrichi en sucre, contenant 75% du régime normal et 25% de saccharose; 3) régime enrichi en graisse de bœuf, contenant 75% du régime normal et 25% de graisse de bœuf; 4) régime enrichi en triglycérides à chaînes moyennes, contenant 75% du régime normal et 25% d'une margarine, contenant 76% de TCM, 4% d'eau, 5% de poudre de lait écrémé, 14% d'eau et 1% de NaCl; 5) régime enrichi en huile de carthame, contenant 75% d'aliment normal et 25% d'huile de carthame; 6) régime enrichi à la fois en sucre et en graisse de bœuf, contenant 50% d'aliment normal, 25% de saccharose et 25% de graisse de bœuf; 7) régime enrichi à la fois en sucre et en TCM, contenant 50% d'aliment normal, 25% de saccharose et 25% de margarine riche en TCM et 8) régime enrichi à la fois en sucre et en huile de carthame, contenant 50% d'aliment normal, 25% de saccharose et 25% d'huile de carthame. En outre, les animaux recevaient comme boisson l'eau des conduites de la ville, mais dans certaines expériences elle contenait 1% de sel.

Méthodes de dosage
1) La glycémie est dosée selon la méthode de *Nelson* [1944]; 2) l'insulinémie est dosée selon la méthode radio-immunologique de *Hales et Randle* [1963]; 3) le cholestérol est

dosé selon la méthode de *Lieberman* [1885] et 4) les triglycérides sont dosés selon la méthode de *Kessler et Lederer* [1965].

L'épreuve d'hyperglycémie
Elle est pratiquée à 9 h du matin sur les animaux mis à jeun depuis la veille à 17 h; ils reçoivent dans une veine de la queue 1 g de glucose par kilogramme. Le sang est prélevé avant, 5, 10, 15 et 20 min après l'injection; la glycémie est dosée sur tous les échantillons; l'insulinémie est dosée avant, 5 et 20 min après l'injection.

Le coefficient d'assimilation du glucose
Il a été déterminé par la formule de *Conard* [1955]

$$\frac{\ln G_1 - \ln G_2}{T_2 - T_1} = K,$$

où G_1 est la glycémie au temps 1 et G_2 la glycémie au temps 2 mesurées en mg/100 ml et T_1 et T_2 le temps 1 et le temps 2 en minutes.

Le coefficient d'efficacité de l'insuline endogène [*Lederer et al.*, 1974]
Il est déterminé en divisant le coefficient d'assimilation du glucose × 100 par la somme de la différence de l'insulinémie en ng/ml aux temps 2 (5 min) et 3 (20 min) par rapport au temps 1 selon la formule

$$\frac{K \times 100}{\Delta \text{insul}_2 + \Delta \text{insul}_3}$$

Le travail physique (nage forcée)
De l'âge de 120 jours à l'âge de 178 jours, des rats des deux sexes, soumis aux différents régimes sont placés chaque matin et chaque après-midi durant 1 h dans un aquarium de 1 m de haut, 40 cm de long et 30 cm de large rempli d'eau à 35 °C jusqu'à 5 cm du bord; ils doivent se débattre énergiquement pour éviter la noyade.

Le taux de cholestérol et le taux de triglycérides ainsi que l'épreuve d'hyperglycémie avec insulinémie sont déterminés la veille de la première séance et le lendemain de la dernière séance de nage forcée.

Les séries
Elles comptent, pour les animaux non soumis à l'exercice physique, de 20 à 36 rats de chaque sexe. Les séries où l'animal fait de l'exercice comptent chacune 12 rats.

La castration
Elle est faite à l'âge de 90 jours; tous les animaux présentant la moindre suppuration sont écartés.

Résultats

Influence des différentes graisses sur la courbe de glycémie et d'insulinémie
L'expérience a été menée sur 8 séries, 4 pour chaque sexe, comptant 30–36 animaux, soumis au régime normal et aux régimes enrichis respective-

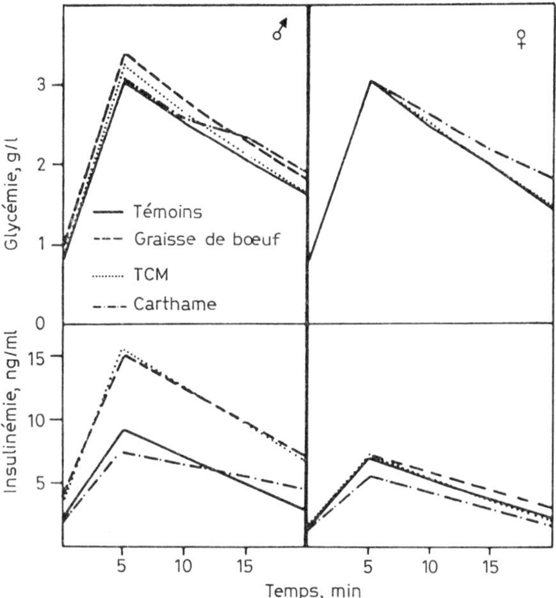

Fig. 1. Courbes de glycémie et d'insulinémie du rat des deux sexes soumis à des régimes enrichis en diverses graisses.

ment en graisse de bœuf, en TCM et en huile de carthame, de l'âge de 120 jours à l'âge de 150 jours.

La courbe de glycémie. Il faut noter chez les mâles comme chez les femelles que la régression de la glycémie est plus lente lorsqu'ils ont reçu un régime enrichi en huile de carthame que lorsqu'ils ont reçu un des trois autres régimes (fig. 1).

Coefficient d'assimilation du glucose. L'ensemble des résultats colligés dans la figure 2 montre que:

1) chez les animaux au régime standard, le coefficient d'assimilation du glucose est supérieur chez les femelles par rapport aux mâles puisqu'il est respectivement de 4,90 chez les femelles et 4,08 chez les mâles;

2) que le régime riche en TCM améliore légèrement le coefficient d'assimilation glucidique chez les mâles puisqu'il le fait passer de 4,08 à 4,62 mais n'a guère influence chez les femelles puisqu'il le fait passer de 4,90 à 4,73;

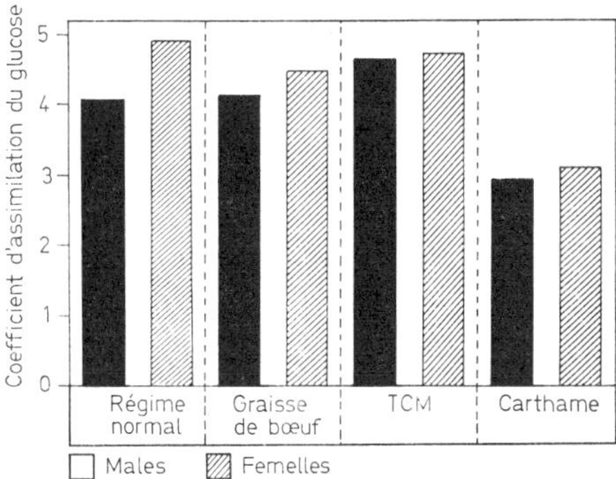

Fig. 2. Coeficient d'assimilation du glucose chez les rats des deux sexes soumis à des régimes enrichis en diverses graisses.

3) que le régime riche en graisse de bœuf ne modifie guère le coefficient d'assimilation glucidique des mâles puisqu'il le fait passer de 4,08 à 4,13 mais qu'il diminue légèrement le coefficient d'assimilation glucidique des femelles puisqu'il passe de 4,90 à 4,45;

4) que le régime riche en huile de carthame provoque une nette diminution du coefficient d'assimilation du glucose puisqu'il le fait tomber chez le mâle de 4,08 à 3,18 et chez la femelle de 4,90 à 3,35.

Il semble donc que les huiles riches en acides gras polydésaturés soient défavorables à l'assimilation glucidique tandis que les graisses saturées le sont moins. Il faut se rappeler à ce sujet que lorsque *Himsworth* [1934] a établi que les régimes gras avaient une action diabétogène chez le lapin, il leur avait administré un régime constitué de 50 g de chou cru et de 120 g de fève de soja, c'est-à-dire un régime riche en graisses polydésaturées.

L'insulino-sécrétion. L'ensemble des résultats obtenus chez les animaux entiers et représentés dans la figure 1 montre que:

1) quel que soit le régime suivi par les animaux, l'insulinémie à jeun est plus élevée chez les mâles que chez les femelles;

2) quel que soit le régime suivi par les animaux femelles, la riposte insulinique à l'injection intraveineuse de glucose est d'à peu près la même importance;

3) chez les animaux mâles, la riposte insulinique est considérablement plus importante lorsqu'ils sont soumis au régime riche en graisse de bœuf

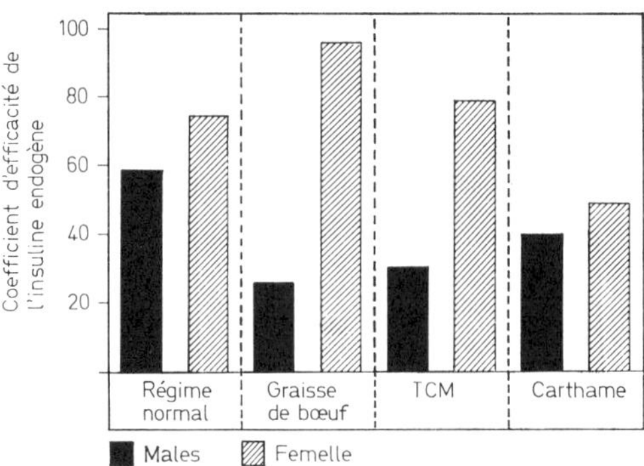

Fig. 3. Coefficient d'efficacité de l'insuline endogène chez les rats des deux sexes soumis à des régimes enrichis en diverses graisses.

ou en TCM que lorsqu'ils sont soumis au régime normal ou au régime riche en huile de carthame;

4) lorsque les animaux sont soumis au régime riche en huile de carthame, la riposte insulinique n'est pas augmentée par rapport aux animaux témoins et cela aussi bien chez les mâles que chez les femelles.

Le coefficient d'efficacité de l'insuline endogène. Chez le mâle, l'enrichissement du régime en graisse provoque une nette diminution de l'efficacité de l'insuline endogène, comme s'il y avait une résistance à l'entrée du glucose dans les cellules malgré une insulinémie élevée.

Chez les femelles, on n'observe une diminution de l'efficacité de l'insuline endogène que lorsqu'elles ont reçu de l'huile de carthame; chez celles qui ont reçu de la graisse de bœuf, il y a une augmentation de l'efficacité de l'insuline endogène et chez celles qui ont reçu des TCM, il n'y a pas de modifications du coefficient d'efficacité de l'insuline endogène (fig. 3).

Influence de l'association du saccharose et des différentes graisses sur l'assimilation du glucose
Cette expérience a été conduite sur 16 séries d'animaux, 8 de rats mâles et 8 de rats femelles soumis de l'âge de 120 à 150 jours à chacun des 8 régimes expérimentaux.

Chaque série comporte de 30 à 40 animaux.

1) L'adjonction de saccharose au régime dans les conditions de notre

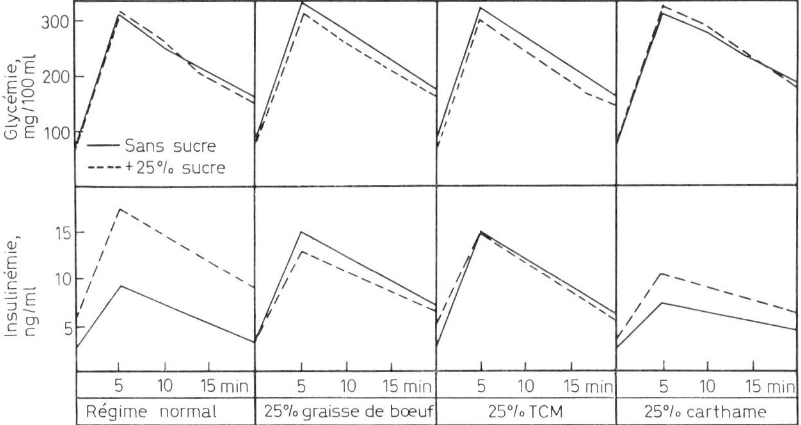

Fig. 4. Courbe de glycémie et d'insulinémie du rat mâle soumis à des régimes contenant à la fois du saccharose et diverses graisses.

expérience n'a modifié la courbe de glycémie ni chez le mâle, ni chez la femelle.

2) Chez les mâles soumis au régime enrichi en graisse de bœuf ou en TCM, l'addition de saccharose ne modifiait pas la riposte insulinique à l'occasion d'une surcharge glucosée; par contre, chez ceux qui étaient soumis au régime normal ou au régime enrichi en huile de carthame, la surcharge glucosée provoquait une augmentation de la riposte insulinique; cette augmentation est d'une moins grande amplitude chez ceux qui reçoivent de l'huile de carthame (fig. 4).

3) Chez les femelles, l'addition de saccharose à tous les régimes provoque une décharge d'insuline à la fois plus importante et plus prolongée; toutefois, c'est avec le régime contenant de l'huile de carthame que l'augmentation est la moins forte (fig. 5).

4) Le coefficient d'assimilation du glucose n'est modifié par l'addition du saccharose ni chez les animaux recevant comme base le régime normal ni chez ceux recevant comme base le régime enrichi en une des trois catégories de graisses; ceci vaut tant pour les mâles que pour les femelles (fig. 6).

5) Le coefficient d'efficacité de l'insuline endogène est abaissé par l'addition du saccharose chez les mâles et chez les femelles recevant comme base le régime normal et chez les femelles recevant comme base le régime enrichi en graisse de bœuf ou en TCM. Dans les autres séries, il n'est pas modifié de manière significative (fig. 7).

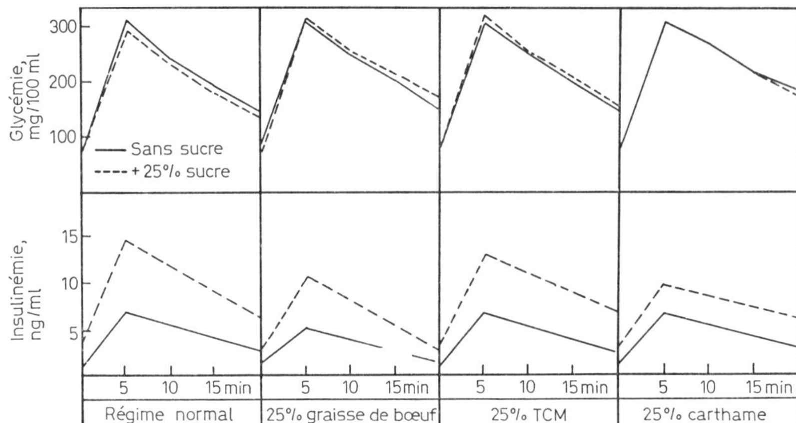

Fig. 5. Courbe de glycémie et d'insulinémie du rat femelle soumis à des régimes contenant à la fois du saccharose et diverses graisses.

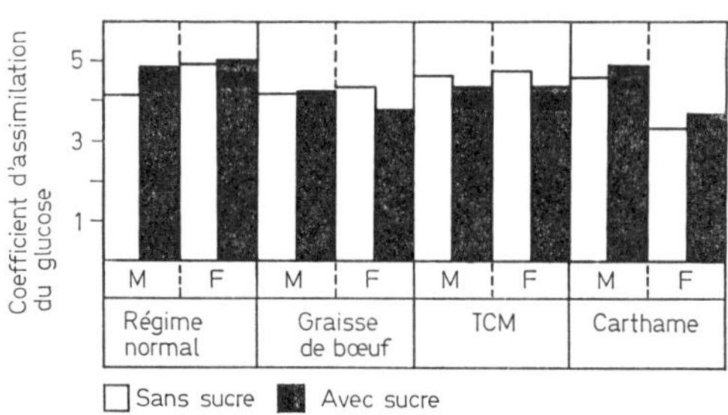

Fig. 6. Coefficient d'assimilation du glucose chez les rats des deux sexes soumis à des régimes enrichis à la fois en saccharose et en diverses graisses.

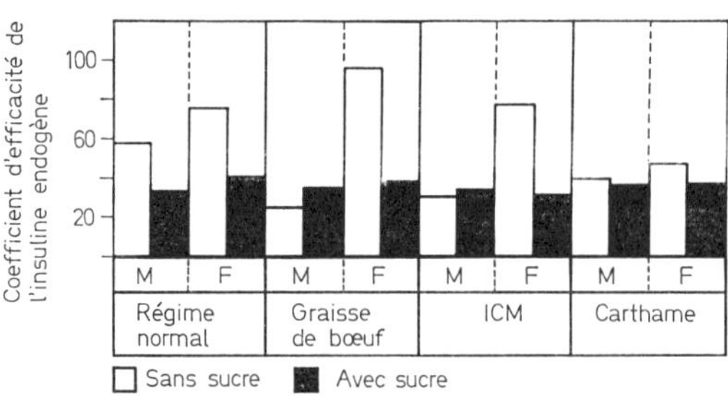

Fig. 7. Coefficient d'efficacité de l'insuline endogène chez le rat des deux sexes soumis à des régimes enrichis à la fois en saccharose et en diverses graisses.

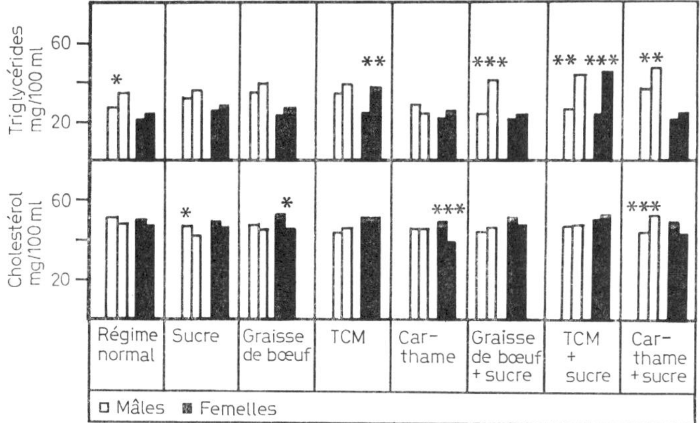

Fig. 8. Influence de régimes chargés en saccharose, en diverses graisses ou à la fois en saccharose et en diverses graisses sur les taux de cholestérol et de triglycérides du rat des deux sexes. *Différences assez significatives; **différences significatives; ***différences très significatives.

Influence respective du sucre et des différentes graisses sur les paramètres lipidiques du rat des deux sexes

Chez les animaux utilisés dans l'expérience précédente, du sang a été recueilli à 120 et à 150 jour de la vie pour y déterminer les taux de cholestérol et de triglycérides.

Les triglycérides subissent des modifications beaucoup plus marquées sous l'influence de ces différents régimes que le cholesterol.

Conformément à ce qu'avait vu *Macdonald* [1972] chez l'homme, l'association graisse saturée-saccharose provoque chez le rat mâle une très forte augmentation des triglycérides plasmatiques alors que ni le saccharose, ni aucune des graisses seules ne provoque d'augmentation de ceux-ci, mais curieusement il y a aussi une augmentation du taux des triglycérides chez eux après le régime associant saccharose et huile de carthame.

Chez la femelle, l'association graisse de bœuf-saccharose ne provoque pas d'hypertriglycéridémie; par contre, les triglycérides à chaînes moyennes seuls ou associés au saccharose provoquent des élévations importantes des triglycérides.

Le cholestérol s'est abaissé légèrement dans les deux sexes sous l'influence de l'administration de saccharose seul; chez la femelle, il s'est abaissé de manière significative avec les régimes contenant de l'huile de carthame

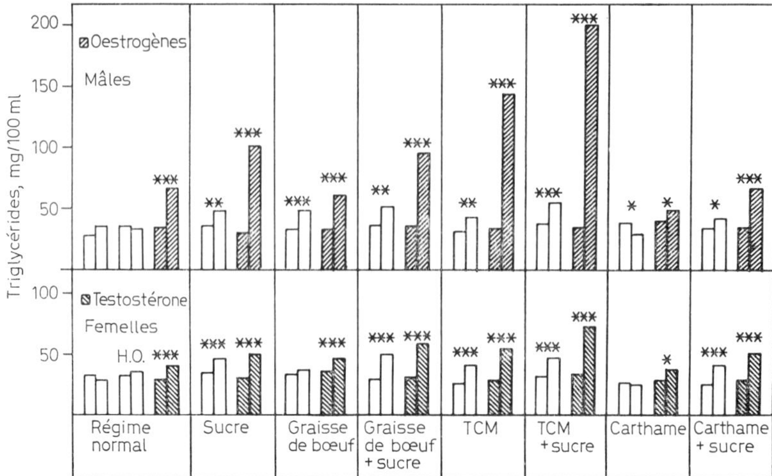

Fig. 9. Influence de divers régimes sur le taux de triglycérides de rats mâles castrés recevant ou non des œstrogènes et de rats femelles castrées recevant ou non des androgènes.

avec ou sans saccharose; il s'est aussi abaissé chez les femelles recevant de la graisse de bœuf. Il ne s'est élevé que dans une seule série, celle des rats mâles recevant à la fois saccharose et huile de carthame (fig. 8).

Influence des hormones sexuelles sur les paramètres lipidiques du rat des deux sexes

Pour vérifier si la différence de comportement des deux sexes vis-à-vis de l'effet lipidogène du saccharose et des différentes graisses était d'origine hormonale ou génétique, nous avons étudié l'effet des huit régimes expérimentaux sur les paramètres lipidiques du rat castré recevant l'hormone sexuelle de l'autre sexe.

Pour cela nous avons mis en expérience 32 séries de 20 animaux tous castrés, soit 16 séries pour chaque sexe avec pour chaque régime une série ne recevant aucun traitement hormonal et une série recevant chaque jour du 120e au 150e jour de la vie 50 μg de benzoate d'œstradiol pour les mâles et 1 mg de propionate de testostérone pour les femelles.

Les triglycérides subissent les différences les plus marquées (fig. 9).

Chez les mâles castrés non traités, le saccharose, la graisse de bœuf, les triglycérides à chaînes moyennes, l'association graisse de bœuf-saccharose et l'association saccharose-triglycérides à chaînes moyennes ont provoqué une augmentation significative mais légère du taux de triglycérides; chez ceux qui avaient reçu de l'huile de carthame, il y a une légère diminution du taux des triglycérides et lorsqu'à l'huile de carthame on ajoute du saccharose, il y a une légère augmentation de ce taux.

Avec tous les régimes, l'administration de benzoate d'œstradiol a provoqué une élévation du taux de triglycérides, même chez les animaux soumis au régime normal; cette élévation est particulièrement importante chez les animaux qui ont reçu du saccharose, des triglycérides à chaînes moyennes, l'association graisse de bœuf-saccharose et surtout l'association saccharose-triglycérides à chaînes moyennes où ce taux quintuple. Par contre, l'élévation est modeste chez les animaux soumis au régime à l'huile de carthame ou à l'association saccharose-huile de carthame.

A noter que chez les animaux soumis au régime normal, l'huile d'olive n'a eu aucune influence sur le taux des triglycérides. En résumé chez le mâle castré, les œstrogènes provoquent une élévation du taux des triglycérides qui est surtout importante chez ceux qui reçoivent soit du saccharose seul soit du saccharose en association avec une graisse saturée; l'huile de carthame, graisse désaturée, inhibe considérablement l'effet du saccharose sur les triglycérides.

Chez la femelle castrée non traitée, le saccharose, les triglycérides à chaînes moyennes, l'association saccharose-graisse de bœuf, l'association saccharose-triglycérides à chaînes moyennes et l'association saccharose-huile de carthame provoquent une élévation du taux des triglycérides tandis que la graisse de bœuf et l'huile de carthame ne provoquent pas de modification.

L'administration de propionate de testostérone accentue de manière très modeste l'élévation du taux de triglycérides dans chacune des séries où on l'avait déjà observée chez les animaux non traités et provoque une légère élévation du taux des triglycérides chez les animaux soumis au régime normal ou enrichi en graisse de bœuf ou en huile de carthame.

L'injection d'huile d'olive ne provoque pas de modifications.

Le cholestérol plasmatique subit des modifications de beaucoup plus faible amplitude que le taux de triglycérides (fig. 10).

Chez le mâle castré ne recevant pas de traitement hormonal, on observe une légère augmentation chez ceux qui ont reçu de la graisse de bœuf-saccharose; en outre, on observe une légère augmentation chez ceux qui, soumis au régime normal, ont reçu des injections d'huile d'olive.

Avec le traitement aux œstrogènes, on observe une nette diminution

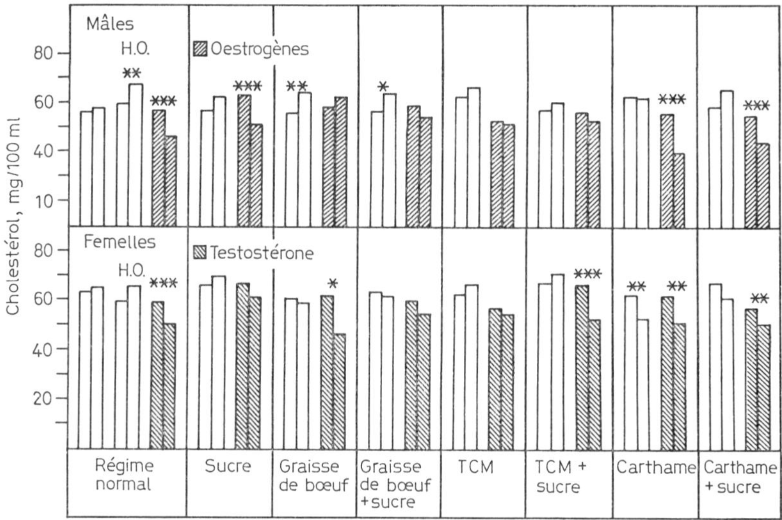

Fig. 10. Influence de divers régimes sur le taux de cholestérol de rats mâles castrés recevant ou non des œstrogènes et de femelles castrées recevant ou non des androgènes.

chez ceux qui reçoivent le régime normal, chargé en saccharose, en huile de carthame ou chargé à la fois en saccharose et en huile de carthame.

Chez les femelles castrées ne recevant pas de traitement hormonal, on observe une diminution du taux de cholestérol chez celles qui reçoivent le régime riche en huile de carthame; il n'y a de modification significative dans aucune des autres séries.

L'administration de propionate de testostérone provoque un abaissement de la cholestérolémie chez celles recevant le régime normal le régime chargé en graisse de bœuf, en saccharose et triglycérides à chaînes moyennes, en saccharose et huile de carthame; elle ne modifie pas l'abaissement observé chez celles recevant l'huile de carthame; elle ne provoque de modification dans aucune autre série.

Donc le benzoate d'œstradiol et le propionate de testostérone favorisent dans l'ensemble un léger abaissement du taux de cholestérol.

Influence des hormones sexuelles sur le métabolisme lipidique des animaux de l'autre sexe

L'hypertriglycéridémie observée chez les rats mâles castrés sous l'effet des œstrogènes surtout lorsqu'ils étaient soumis à un régime enrichi à la fois

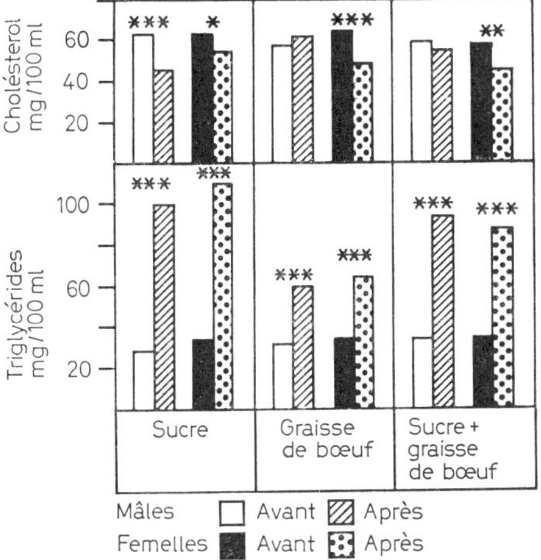

Fig. 11. Influence des œstrogènes sur les taux de cholestérol et de triglycérides du rat castré des deux sexes soumis à un régime enrichi soit en saccharose, soit en graisse de bœuf, soit à la fois en saccharose et en graisse de bœuf.

en sucre et en graisses saturées permettait de se demander si ceci était dû à une sensibilité particulière des mâles, d'origine génétique, ou s'il s'agissait d'un phénomène purement hormonal.

Pour vérifier cela, l'expérience a été recommencée sur des rats femelles castrés auxquels on administrait chaque jour 50 μg de benzoate d'œstradiol de manière à comparer les résultats avec ceux obtenus chez les rats mâles castrés soumis au même traitement.

Trois régimes ont été étudiés; celui enrichi en graisse de bœuf, celui enrichi en saccharose et celui enrichi à la fois en saccharose et en graisse de bœuf.

Il apparaît que l'élévation du taux des triglycérides est de la même ampleur chez les mâles et chez les femelles pour les trois régimes étudiés; il n'y a donc pas de différence de sensibilité à ce point de vue entre les deux sexes.

En ce qui concerne le taux de cholestérol, alors que chez le mâle castré recevant des œstrogènes, il n'y a pas de modification avec le régime enrichi en graisse de bœuf ou enrichi à la fois en graisse de bœuf et en saccharose, chez la femelle on observe une légère diminution (fig. 11).

Etant donné que les animaux mâles sont beaucoup plus sensibles aux facteurs athérogènes, nous avons voulu voir si le rat mâle castré traité par

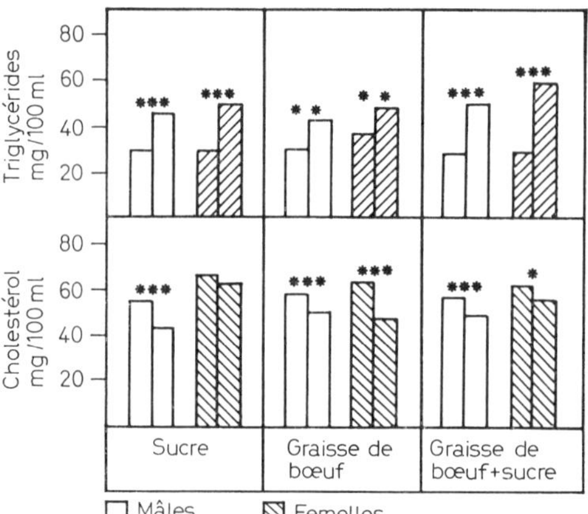

Fig. 12. Influence de la testostérone sur les taux de cholestérol et de triglycérides du rat castré des deux sexes soumis à un régime enrichi soit en saccharose, soit en graisse de bœuf, soit à la fois en saccharose et en graisse de bœuf.

la testostérone était plus sensible que la femelle à l'effet lipidogène d'un régime chargé en graisse saturée et en saccharose.

Pour vérifier cela, nous avons déterminé les modifications du taux de cholestérol et de triglycérides chez des rats mâles castrés recevant chaque jour 1 mg de propionate de testostérone et soumis à un régime enrichi, soit en saccharose, soit en graisse de bœuf, soit à la fois en saccharose et en graisse de bœuf. Les résultats ont été comparés d'une part avec ceux observés chez les rats mâles castrés ne recevant pas de traitement hormonal et d'autre part chez les femelles castrées soumises au même régime, recevant ou ne recevant pas de propionate de testostérone (fig. 12).

Les trois régimes provoquent chez les animaux castrés des deux sexes une élévation du taux des triglycérides de la même amplitude. Le propionate de testostérone ne modifie pas celle-ci chez les mâles et l'augmente légèrement chez les femelles.

Le taux de cholestérol qui augmentait avec les trois régimes chez les mâles castrés non traités diminue légèrement sous l'effet du propionate de testostérone: chez les femelles castrées non traitées, aucun des trois régimes ne provoque de modification significative; sous l'effet du propionate de testostérone, on observe une légère diminution avec le régime enrichi en graisse de bœuf et celui enrichi à la fois en saccharose et en graisse de bœuf.

Au total, on n'observe pas, dans les limites de cette expérience, une

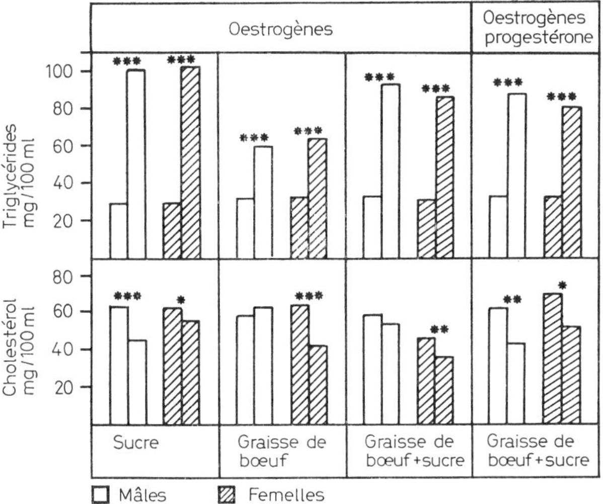

Fig. 13. Influence de la progestérone sur l'hypertriglycéridémie déclenchée par les œstrogènes chez le rat castré des deux sexes soumis à un régime enrichi à la fois en graisse de bœuf et en saccharose.

différence de sensibilité du métabolisme lipidique au propionate de testostérone.

La progestérone protège-t-elle contre l'effet lipidogène des œstrogènes?

Pour vérifier si la raison pour laquelle les femelles recevant un régime enrichi en graisse de bœuf et en saccharose ne font pas d'hypertriglycéridémie n'était pas la sécrétion de progestérone puisque les œstrogènes la favorisent, nous avons fait une nouvelle expérience consistant à comparer l'effet de la progestérone sur l'hypertriglycéridémie provoquée par les œstrogènes aussi bien chez les mâles que chez les femelles castrés soumis au régime enrichi à la fois en saccharose et en graisse de bœuf. Chaque série compte 12 animaux.

Des rats mâles et des rats femelles castrés reçoivent un régime enrichi en graisse de bœuf et en saccharose du 120e au 150e jour, tous reçoivent chaque jour une injection de 0,1 mg de benzoate d'œstradiol et une moitié d'entre eux reçoit, en outre, chaque jour 1 mg de progestérone.

L'administration de progestérone n'a pu atténuer l'hypertriglycéridémie favorisée par les œstrogènes ni chez les mâles, ni chez les femelles; elle n'a pas d'avantage modifié son influence sur le taux de cholestérol (fig. 13).

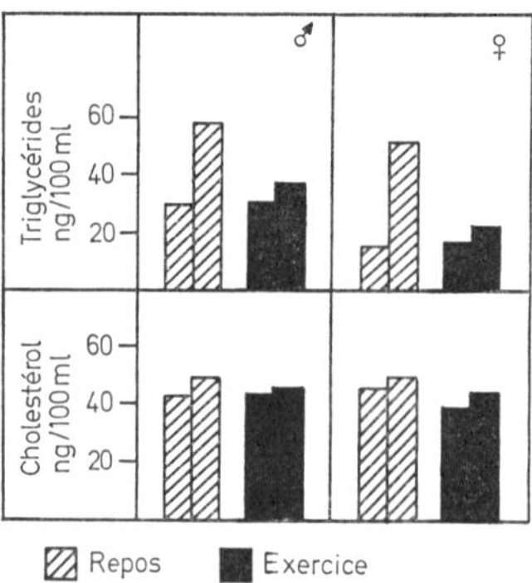

Fig. 14. Influence de l'exercice physique sur les modifications du taux de cholestérol et de triglycérides chez le rat des deux sexes soumis à un régime enrichi à la fois en graisse de bœuf et en saccharose.

Influence de l'exercice physique sur l'action lipidogène du régime chargé en graisse saturée et en sucre

Etant donné le rôle de la sédentarité dans la pathogénie de l'athéromatose, nous avons voulu voir quelle était l'influence de l'exercice physique sur l'effet d'un régime lipidogène.

Des rats mâles et des rats femelles ont été soumis de l'âge de 120 jours à l'âge de 177 jours à un régime enrichi à la fois en graisse de bœuf et en saccharose; une moitié d'entre eux restait en cage et l'autre moitié était astreinte deux fois par jour à 1 h de nage forcée. Chaque série comptait 12 animaux. Les taux de cholestérol et de triglycérides sont déterminés le premier et le dernier jour de cette expérience.

Chez les animaux restés au repos, il y a une forte augmentation du taux des triglycérides; celle-ci est fortement atténuée et non significative chez ceux qui ont été astreints à la nage forcée. Le taux de cholestérol n'a pas été modifié de manière significative (fig. 14).

Des expériences avec des temps de travail plus courts nous ont montré que pour avoir un effet il fallait faire nager les rats au moins deux fois par jour durant 45 min.

L'exercice physique est donc capable de combattre l'effet d'un régime lipidogène sur le taux de triglycérides à condition d'être intense et prolongé; il est aussi capable d'atténuer la décharge d'insuline provoquée par une injection intraveineuse de glucose, même quand celle-ci est faite après une nuit de repos (fig. 15).

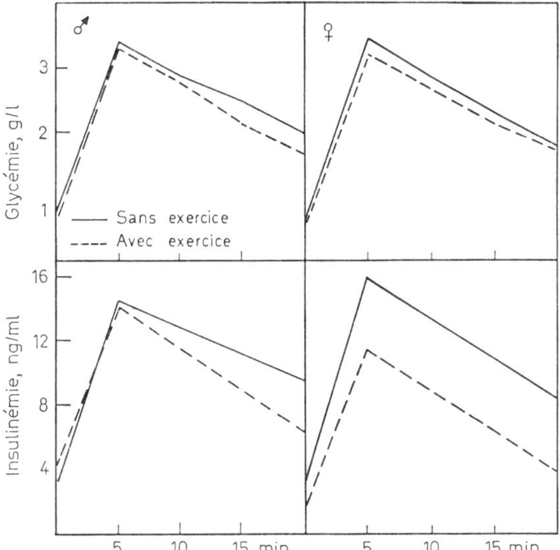

Fig. 15. Influence de l'exercice physique sur la courbe de glycémie et d'insulinémie du rat des deux sexes soumis à un régime enrichi en graisse de bœuf et en saccharose.

Influence du chlorure de sodium sur les effets métaboliques de régimes chargés en graisse et en sucre

Etant donné le rôle néfaste du sel dans la régulation de la tension artérielle et par là dans la pathogénie de l'athéromatose, nous avons voulu voir s'il n'avait pas aussi une influence sur les effets métaboliques de régimes chargés à la fois en graisse et en sucre.

Pour étudier ce point, nous avons soumis des rats mâles et des rats femelles de l'âge de 120 jours à l'âge de 150 jours à six régimes différents: régime normal, régime chargé en saccharose, régime chargé en graisse de bœuf, régime chargé en huile de carthame, régime chargé à la fois en graisse de bœuf et en saccharose, régime chargé à la fois en huile de carthame et en saccharose.

Une moitié des animaux recevait comme boisson de l'eau pure, l'autre moitié recevait de l'eau contenant 1% de chlorure de sodium.

Les modifications du taux de cholestérol ont été dans l'ensemble peu importantes. Les seules différences vraiment significatives en ont été une chute chez les rats femelles buvant l'eau salée et mises au régime enrichi en

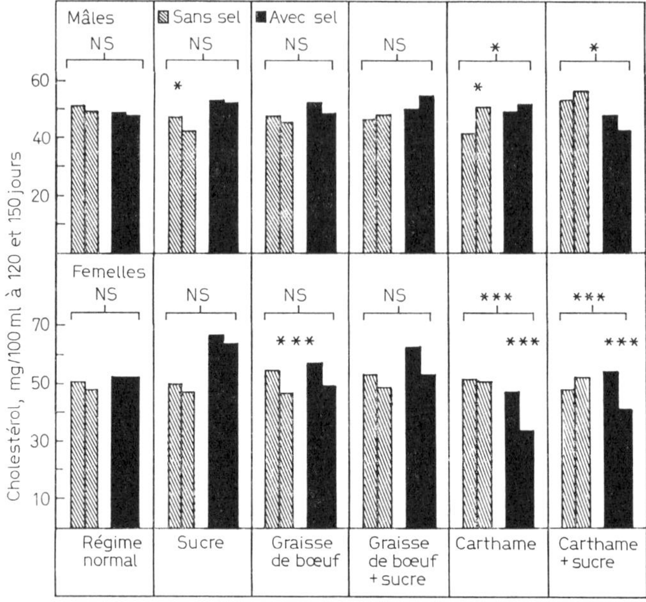

Fig. 16. Influence du chlorure de sodium sur les modifications du taux de cholestérol chez le rat des deux sexes soumis à différents régimes.

huile de carthame et au régime enrichi à la fois en huile de carthame et en saccharose (fig. 16).

Le taux de triglycérides est augmenté fortement chez les rats mâles au régime enrichi en sucre et buvant l'eau salée de même que chez les femelles au régime enrichi à la fois en graisse de bœuf et en saccharose et buvant l'eau salée. Ce sont les deux seules séries où l'on voit une influence du sel sur l'action lipidogène d'un régime (fig. 17).

On observe aussi une augmentation du taux des triglycérides chez les mâles consommant un régime enrichi à la fois en graisse de bœuf et en saccharose, mais elle est de la même importance chez ceux qui boivent de l'eau pure et qui boivent de l'eau salée; on observe une diminution du taux des triglycérides chez les rats mâles qui sont au régime chargé à la fois en huile de carthame et en saccharose mais elle est de la même importance avec l'eau pure et avec l'eau salée.

En outre, nous avons constaté que l'addition de sel à l'eau de boisson provoque chez les animaux mâles et pas chez les femelles une augmentation

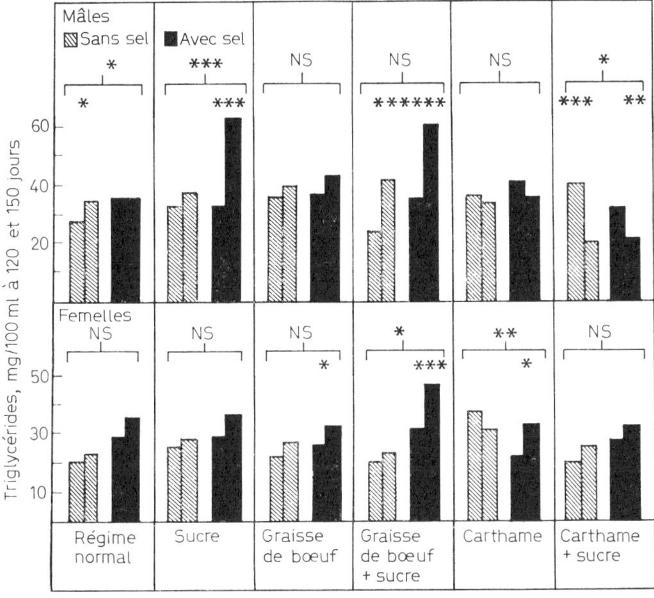

Fig. 17. Influence du chlorure de sodium sur les modifications du taux de triglycérides chez le rat des deux sexes soumis à différents régimes.

de la décharge d'insuline à l'occasion d'une surcharge glucosée, sans modifier la courbe de glycémie (fig. 18).

Discussion

L'ensemble des recherches que nous avons entreprises nous a montré l'importance de l'acide linoléique dans l'équilibre de la ration alimentaire.

Le sucre, dont *Yudkin* [1957] avait dénoncé la nocivité ne provoque une hypertriglycéridémie que s'il est consommé avec des graisses saturées et encore chez les rats mâles uniquement.

Nous confirmons chez l'animal d'expérience les constatations qu'avait faites *Macdonald* [1972] chez l'homme.

Il faut noter à ce propos que l'action lipidogène des graisses saturées est particulièrement manifeste pour les triglycérides à chaînes moyennes puisqu'ils provoquent de l'hypertriglycéridémie même chez les femelles et

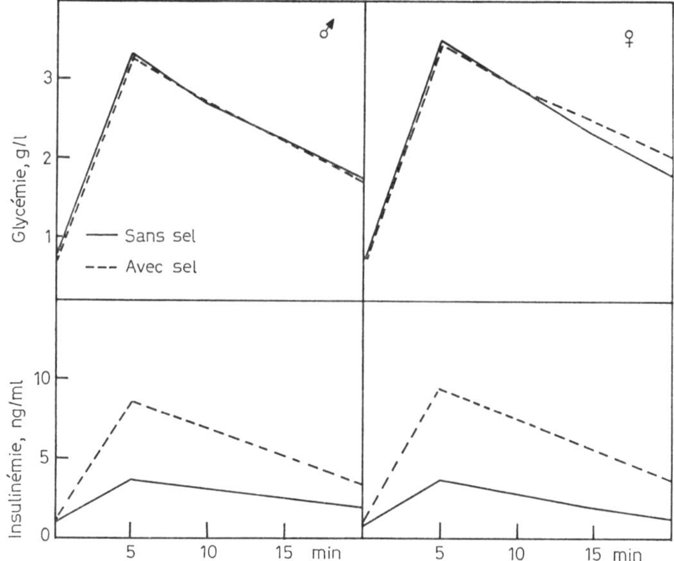

Fig. 18. Influence du chlorure de sodium sur la courbe de glycémie et d'insulinémie du rat des deux sexes.

qu'en association avec du saccharose ils provoquent une hypertriglycéridémie plus importante que la graisse de bœuf.

La nocivité des graisses saturées et l'effet protecteur de l'acide linoléique ne sont pas seulement liés à leur influence sur le taux de triglycérides et de cholestérol mais aussi à leur répercussion sur la riposte insulinique à l'occasion d'une surcharge glucosée.

Nous avons observé en effet que, du moins chez le rat mâle soumis à un régime enrichi en graisse de bœuf ou en triglycérides à chaînes moyennes, la décharge d'insuline est beaucoup plus forte que lorsqu'il est soumis à un régime pauvre en graisse ou contenant des graisses riches en acide linoléique.

Ceci ne se voit pas chez la femelle.

Ce fait est fort important; on sait en effet par les travaux de *Stamler* [1960], de *Renold et al.* [1965], de *Stout* [1968] que l'hyperinsulinisme favorise l'anabolisme lipidique au niveau de la paroi de l'artère et freine le catabolisme.

Il faut noter que lorsqu'aux graisses de quelque nature qu'elles soient on ajoute du saccharose, la riposte insulinique à l'occasion d'une surcharge glucosée atteint le niveau de celle que l'on observe chez le rat mâle soumis au régime enrichi en graisse de bœuf et ceci même chez la femelle.

De toute façon, un régime enrichi à la fois en saccharose et en graisses, surtout si elles sont saturées et surtout chez les mâles, est nocif. Ceci explique la constatation de *Kohn* [1973] qui avait observé que l'homme grand amateur de chocolat meurt trois fois plus souvent d'infarctus que l'homme non-amateur de chocolat.

La richesse du régime en chlorure de sodium intervient aussi dans la pathogénie de l'hypertriglycéridémie; l'enrichissement en sel provoque une hypertriglycéridémie nette surtout chez le rat mâle soumis au régime riche en sucre ou chez la femelle soumise au régime riche à la fois en sucre et en graisse de bœuf; lorsque le sel est donné dans un régime riche en acide linoléique avec ou sans sucre il ne provoque plus d'hypertriglycéridémie et même peut la faire baisser, surtout chez le mâle. L'effet nocif du sel est dû encore au fait qu'il provoque outre l'hypertension de l'hyperinsulinisme.

L'effet nocif de l'association graisse saturée-saccharose peut être combattu par l'exercice physique; nous avons pu diminuer considérablement; tant chez les rats mâles que chez les femelles, l'hypertriglycéridémie provoquée par un régime enrichi à la fois en graisse de bœuf et en saccharose en les astreignant à un exercice intense de nage forcée 2 h par jour.

Les recherches que nous avons poursuivies pour tenter de trouver une explication à la différence de réaction des mâles et des femelles ne nous ont pas donné de réponse satisfaisante.

Les facteurs endocriniens ne semblent pas être en cause puisque l'administration d'œstrogènes chez le rat mâle castré soumis au régime enrichi en graisse de bœuf et saccharose provoque une hypertriglycéridémie plus forte que l'administration d'androgènes à la femelle soumise au même régime.

Contrairement à *Kim et Kulkhoff* [1975], nous n'avons pas pu empêcher par la progestérone l'action des œstrogènes sur les triglycérides, mais ces auteurs avaient administré une dose de progestérone 1000 fois supérieure à celle du benzoate d'œstradiol tandis que nous nous sommes placés dans les conditions physiologiques en administrant seulement 10 fois plus de progestérone que de benzoate d'œstradiol.

Nous n'avons pas pu d'avantage mettre en évidence un facteur génétique puisque chez la femelle castrée et chez le mâle castré soumis aux même régimes, les œstrogènes provoquent une hypertriglycéridémie de la même importance; il en va de même des androgènes.

Résumé

L'acide linoléique joue un rôle capital dans l'équilibre de la ration alimentaire chez le rat car: 1) chez le rat mâle un régime riche en graisse de bœuf favorise l'hyperinsulinisme à l'occasion d'une surcharge glucosée tandis que les graisses riches en acides linoléiques ne le font pas; 2) l'association graisse saturée-saccharose favorise surtout chez le rat mâle de l'hypertriglycéridémie, ce que ne font pas les associations graisses riches en acide linoléique-saccharose; 3) l'acide linoléique empêche le chlorure de sodium de provoquer de l'hyper-triglycéridémie chez le rat des deux sexes soumis à un régime enrichi à la fois en saccharose et en graisse; 4) l'hypertriglycéridémie provoquée par les œstrogènes chez les animaux castrés est fortement diminuée si le régime est riche en acide linoléique; 5) l'effort physique peut empêcher l'association graisse saturée-saccharose de provoquer de l'hypertriglycéri-démie. En montrant que la richesse en acide linoléique d'une graisse influençait de manière décisive l'action lipidogène du fructose, *Macdonald* [1972] a ouvert une voie nouvelle pour les recherches dans le domaine de la pathogénie de l'athéromatose.

Bibliographie

Conard, V.: Mesure de l'assimilation du glucose. Bases théoriques et applications cliniques. Acta gastro-enter. belg. *18:* 803–812 (1955).

Hales, C.N. and Randle, P.J.: Immunoassay of insulin with insulin-antibody precipitates. Biochem. J. *88:* 137–146 (1963).

Himsworth, H.P.: Dietetic factors influencing glucose tolerance and activity of insulin. J. Physiol., Lond. *81:* 29–48 (1934).

Kessler, G. and Lederer, H.: Fluorometric measurement of triglycerides. Autom. analyt. Chem. *1965:* 341–344.

Keys, A.; Anderson, J.T., and Grande, F.: Essential fatty acids degree of unsaturation and effect of corn (maize) oil on the serum level of cholesterol in man. Lancet *i:* 66–68 (1957).

Kim, H.J. and Kalkhoff, R.: Sex steroids influence on triglyceride metabolism. J. clin. Invest. *56:* 888–896 (1975).

Kinsell, L.W.; Michaels, G.D.; Friskey, R.W.; Brown, F.R., Jr., and Maryana, F.: Essential fatty acids, lipid transport and degenerative vascular disease. Circulation *14:* 484–492 (1956).

Kohn, L.A.: Chocolate and coronary heart disease. Am. J. clin. Nutr. *23:* 2–3 (1970).

Lederer, J.; Pottier-Arnould, A.M.; Niethals, E. et Godts Sevens, C.: Influence des graisses sur l'assimilation glucidique chez le rat des deux sexes. Bull. Acad. r. Méd. Belg. *129:* 547–572 (1974).

Lieberman, C.: Über das Oxynoterpen. Ber. dt. chem. Ges. *18:* 1303–1305 (1885).

Macdonald, I.: Relationship between dietary carbohydrates and fats in their influence on serum lipid concentrations. Clin. Sci. *43:* 265–274 (1972).

Nelson, N.: A photometric adaptation of the Somogyi method for the determination of glucose. J. physiol. Chem. *153:* 375–380 (1944).

Renold, A.E.; Crofford, O.B.; Stauffacher, W., and Jeanrenaud, B.: Hormonal control of adipose tissue metabolism with special reference to the effects of insulin. Diabetologia *1:* 4–12 (1965).
Stamler, J.; Pick, R., and Katz, N.: Effect of insulin in the induction and regression of atherosclerosis in the chick. Circulation Res. *8:* 572–576 (1960).
Stout, R.W.: Insulin stimulated lipogenesis in arterial tissue in relation to diabetes and atheroma. Lancet *ii:* 702–703 (1968).
Stout, R.W.: Insulin stimulation of cholesterol synthesis by arterial tissue. Lancet *i:* 248 (1970).
Yudkin, D.: Diet and coronary trombosis. Hypothesis and fact. Lancet *ii:* 155–162 (1957).

J. Lederer, Université de Louvain, Ecole de Santé Publique, Unité de Nutrition, 30, avenue Chapelle-aux-Champs, B-1200 Bruxelles (Belgique)

Nutr. Metab. *24* (Suppl. 1): 142–146 (1980)

Pathophysiology of Long-Chain Polyene Fatty Acids in Heart Muscle

Sigmundur Gudbjarnason

Science Institute, University of Iceland, Reykjavik

Key Words. Heart muscle · Phospholipids · Polyene fatty acids · Aging · Stress · Diet · Docosahexaenoic acid · Sudden cardiac death · Atherosclerosis

Abstract. The polyene fatty acid composition of cardiac phospholipids is modified by age, diet and stress in the rat. In the aging heart there is a progressive replacement of 18:2n6 by 20:4n6 in phosphatidylcholine (PC) and a replacement of 18:2n6 by 22:6n3 in phosphatidylethanolamine (PE). Norepinephrine stress accelerates aging of cardiac PC and PE. Dietary fish oil causes a replacement of 18:2n6 and 20:4n6 by 22:6n3 in cardiac PC and PE but not in cardiolipin. Studies on human cardiac autopsy samples suggest that: (a) polyene fatty acid composition changes with age; (b) stability of cardiac phospholipids is a function of the fatty acid composition, chain length and unsaturation; (c) coronary atherosclerosis is associated with a reduced content of 18:2n6 in phospholipids, an increased content of glycerides of abnormal composition and an unexpectedly low level of free fatty acids (FFA) in the heart muscle, and (d) many cases of sudden cardiac death in the absence of marked coronary artery stenosis or myocardial infarction may be associated with significant alterations in myocardial levels of FFA ($\uparrow$) or PE ($\downarrow$).

The purpose of this study was to examine the role of long-chain polyene fatty acids in cardiac metabolism and function. Polyene fatty acids in heart muscle are mainly present in the phospholipids. The phospholipids play an important role in the structure and function of membranes and these membranes are dynamic structures in a state of constant activity. In the heart muscle, the polyene fatty acyl groups of phospholipids are modified in response to variations in demands upon the organ and during adaptation to external or internal changes.

In the present paper, we examine changes in fatty acid composition of cardiac lipids in relation to age, stress and dietary lipids in the rat, and in relation to age, coronary atherosclerosis and sudden cardiac death in man.

I. Modification of Polyene Fatty Acids in Phospholipids of Rat Heart Muscle

Age and Cardiac Phospholipids

The long-chain polyene fatty acid composition of cardiac phospholipids changed markedly with age in rats. In phosphatidylcholine (PC) there was a progressive replacement of linoleic acid, 18:2n6, by arachidonic acid, 20:4n6, with age. Linoleic acid decreased from 28.4% in 2-month-old rats to 17% in 18-month-old animals. Arachidonic acid increased from 11.7 to 20.6% during the same period. In phosphatidylethanolamine (PE), the level of docosahexaenoic acid, 22:6n3, was high or 20.1% at age 2 months and increased to 28.9% at age 18 months, replacing 18:2n6, which decreased from 15.0 to 8.7% during this period. The relative content of 20:4n6 remained constant during this time. The relative contents of saturated and monoene fatty acids in cardiac phospholipids did not significantly change during this period.

Stress and Cardiac Phospholipids

Rats were subjected to catecholamine stress, i.e. repeated administration of norepinephrine. The animals were daily injected s.c. with norepinephrine, the first 3 days they received 1 mg/kg and 2,3,4 and 5 mg/kg for the subsequent 3-day periods. Norepinephrine stress significantly altered the polyene fatty composition of PC and PE. During the 15 days of norepinephrine administration, cardiac phospholipids in 6-month-old rats had 'aged' by more than 1 year and the fatty acid composition of PC and PE corresponded to that observed in old animals (>2 years). In PC, the 18:2n6 decreased from 22.4 to 11.1% and 20:4n6 and 22:6n3 significantly increased during the period of stress. In PE, we observed an increased in 22:6n3, from 22.9 to 28.9%, and a significant decrease in 20:4n6 during norepinephrine administration. The mortality during norepinephrine stress was about 50%.

Dietary Lipids and Cardiac Phospholipids

Dietary fat is not only involved in hyperlipemia and development of atherosclerosis, but it may also cause extensive alterations in the composition of cardiac lipids and thereby influence membrane properties. Rats fed a diet containing 10% cod liver showed significant alterations in fatty acid composition of cardiac glycerides and phospholipids.

In cardiac glycerides, there was a significant decrease in 16:0, 18:2n6 and 20:4n6 and replacement of these fatty acids by 20:1n9, 20:5n3, 22:1n11 and 22:6n3 in animals fed cod liver oil. In cardiac phospholipids, there was a significant decrease in 18:2n6 and 20:4n6 and replacement by 22:6n3. The dietary fish oil caused significant changes in the fatty acid composition of PC and PE, but the composition of cardiolipin, which contained about 70% 18:2n6, was not markedly altered by the diet.

II. Studies on Human Heart Muscle

Analysis of human cardiac autopsy samples suggested that changes in tissue levels of free fatty acids and/or polyene fatty acids in specific phospholipids may be associated with coronary heart disease. Heart muscle samples were obtained from people who died a sudden accidental death and from people who died a sudden cardiac death with or without coronary atherosclerosis. Sudden cardiac death is defined as instant death or death within 1 h, usually within minutes, of symptoms of myocardial infarction.

Polyene Fatty Acids and the Aging Heart

The fatty acid composition of human cardiac phospholipids changed with age in a manner similar to the alterations observed in rats. There was a significant diminution in the linoleic acid content of cardiac phospholipids in the aging heart accompanied by an increase in docosahexaenoic acid, whereas 20:4n6 and the saturated and monoene fatty acids remained unchanged. This decrease in 18:2n6 with age may be related to development of coronary atherosclerosis since there was a markedly lower content of 18:2n6 in normal cardiac muscle in the presence of severe coronary atherosclerosis.

Changes in the composition of membrane lipids are of considerable interest since age-induced changes in the heart are likely to affect its susceptibility to damage and its response to drugs. One of the most striking physiological changes known to occur with age is a decline in the ability of the heart to respond to stress.

Stability of Cardiac Phospholipids

The composition of free fatty acids (FFA) found in autopsy samples indicates that these FFA were primarily derived from the hydrolysis of cardiac phospholipids. There was an important difference in the composition of phospholipids and FFA, the FFA contained relatively more of the polyene fatty acids, 20:4n6 and 22:6n3, than the remaining phospholipids. This could indicate that phospholipids containing these fatty acids were more readily hydrolyzed by myocardial phospholipases than phospholipids containing more saturated fatty acids. There also appeared to be a faster breakdown of phospholipids containing 22:6n3 than of those containing 20:4n6. There was a specific relationship between 20:4n6 and 22:6n3 in cardiac PC and PE, respectively The relationship between these two fatty acids in myocardial FFA, mostly released from the phospholipids, showed that there is relatively more 22:6n3 than 20:4n6 in the FFA fraction compared to remaining phospholipids (PC and PE), suggesting that phospholipids containing 22:6n3 break down more readily than other phospholipids.

These observations indicate that the stability of human cardiac phospholipids is a function of the fatty acid composition. The most unsaturated phospholipids are most unstable and the resistance to hydrolysis increases with the saturation of the fatty acids. Conditions which modify the fatty acid composition of phospholipids may thus influence the stability of membrane phospholipids and thereby affect cellular susceptibility to damage.

Myocardial Lipids and Sudden Cardiac Death

A. In sudden cardiac death with no or mild coronary atherosclerosis the following cases were observed:

(a) Very high myocardial levels of FFA (14.2 ± 2.3 mg/g), derived from glycerides, normal phospholipids: excessive FFA levels are known to impair myocardial energy production and energy transfer.

(b) Elevated myocardial levels of FFA (6.2 ± 0.2 mg/g), derived from phospholipids and glycerides: myocardial levels of PE were low and had very low levels of 22:6n3. Cellular membrane damage prior to death possibly accompanied by impaired Na^+ conduction is suggested.

(c) Normal levels of myocardial FFA (4.3 ± 0.3 mg/g), abnormal glyceride composition.

B. In sudden cardiac death with marked coronary artery stenosis, grades 5 or 6, the following was observed:

(a) Low levels of FFA (1.8 ± 0.5 mg/g).

(b) Cardiac glycerides were frequently increased in quantity in atherosclerotic hearts. The unsaturation of glyceride fatty acids increased with the glyceride content of the muscle, suggesting that the pathological glycerides may be derived from phospholipids by the action of phospholipase C in an ischemic or energy-deficient muscle.

(c) Phospholipids from normal heart muscle with severe coronary atherosclerosis had a significantly lower content of 18:2n6 compared to normal muscle of hearts with no or mild coronary atherosclerosis.

(d) These individuals frequently had myocardial infarction.

These observations suggest that significant alterations in cardiac lipid composition or tissue content may be associated with coronary artery disease and many cases of sudden cardiac death in the absence of coronary artery stenosis or myocardinal infarction. Sudden cardiac death may be associated with: (a) myocardial energy deficiency due to ischemia and infarction, (b) disturbances in myocardial energy production or energy transfer due to excessive cellular levels of FFA, (c) excessive breakdown of PE and membrane damage and (d) other causes.

Prof. S. Gudbjarnason, Department of Chemistry, Science Institute,
University of Iceland, Dunhaga 3, 107 Reykjavik (Iceland)

Discussion

Prof. Epstein: Do the age differences explain the differences in the fatty acid composition?

Prof. Gudbjarnason: The decrease in linoleic acid with advancing age may be related to the development of atherosclerosis. When we compare the normal cardiac muscle from hearts with no or mild atherosclerosis to muscle samples from hearts with severe coronary stenosis we find significantly lower levels of linoleic acid in the atherosclerotic hearts.

Nutr. Metab. *24* (Suppl. 1): 147–161 (1980)

The Role of Fat in Child Nutrition

Francisco Grande

Institute of Biochemistry and Nutrition, Fundación F. Cuenca Villoro,
and Department of Biochemistry, University of Zaragoza, Zaragoza

Key Words. Fat · Cow's milk · Human milk · Fatty acids · Essential fatty acids ·
Essential fatty acid requirements · Glyceride structure · Premature infants ·
Serum cholesterol · Dietary fat and serum cholesterol

Abstract. The wide differences in fat content among the milks produced by different
mammals seem to indicate that the fat requirements of the newborn are also different. Fat
accounts for over 50% of the energy content of human milk. This relatively high fat content can be necessary to meet the baby's high energy demands during the first weeks of
life, when it has a reduced ability to adjust the volume of intake.

Data from breast-fed babies showing satisfactory growth indicate that their fat intake
up to 5–6 months of age is of the order of 34–45 g/day.

Digestion and absorption of fat are influenced by the degree of saturation and molecular size of the fatty acids, and by the triglyceride structure. In general, fats rich in
unsaturated fatty acids are better absorbed than saturated fats, both by children and adults.
Short-chain fatty acids are better absorbed than those of greater molecular size, for an
equal degree of saturation.

Fats having most of their palmitic acid esterfied in position 2 are better absorbed
than fats having this fatty acid randomly distributed among the three positions of the
glyceride. The better absorption of human milk fat as compared to that of cow's milk
can be explained, at least in part, by the preferential esterification of palmitic acid in
position 2 of the glycerides of human milk fat.

Difficulties with fat absorption are particularly important in premature babies. These
babies are able to absorb human milk fat to the extent of 80%, but have a smaller ability
to absorb the fat from cow's milk.

Human milk has been shown to contain a lipase which is activated by bile salts and
seems to be important for the digestion of milk lipids, particularly the esters of retinol.

Current estimates of essential fatty acid (EFA) requirements by infants and children
are of the order of 3% of the total energy intake. This figure is lower than the EFA content
of human milk generally reported, but higher than that of cow's milk. There is no evidence
that infants fed solely on cow's milk during their first 4–5 months have clinical manifestations of EFA deficiency. However, bottle-fed babies tend to have lower EFA levels and
higher ratios of trienes to tetraenes in their blood lipids, than breast-fed babies.

Plasma cholesterol levels of infants and children are affected by the fatty acid composition of the dietary fat and the plasma cholesterol of brest-fed babies may be reduced by increasing the linoleic acid content of the mother's milk. This role of fat is of obvious interest in child nutrition, in view of the emphasis currently laid on the early prevention of atherosclerosis. An analysis of this problem is, however, beyond the scope of this discussion.

It is generally accepted that dietary fat fulfills the following nutritional functions as (1) a source of energy; (2) a source of building materials for various cellular structures; (3) a source of essential fatty acids, and (4) as a factor in the control of plasma lipid levels. In addition, dietary fat is an important contributor to the palatability of food, and is of paramount importance in cooking and preservation of food. These aspects of the nutritional role of fat are, obviously, more important in adult nutrition than in the feeding of infants, and will not be considered in our discussion.

Fat as a Source of Energy in Infancy

The wide differences in fat content of the milks produced by different species of mammals supports the view that the fat requirements of their newborns may also be widely different. Thus, the fat content of the Grey Seal's milk *(Halycrocus grypus)* is 53.2 g/dl and that of the Blue Whale's *(Balaenoptera musculus)* is 42.3 g/dl. In these two species producing milks of high energy value, fat represents 90 and 88% of the total milk energy.

At the other extreme, donkey's milk *(Equus asinus)* has a fat content of 1.4 g/dl or 25% of its total energy.

The milk of the rhinoceros *(Diceros bicornis)* is reported to be practically fat free [58].

The fat content of human milk is of the order of 4–4.5 g/dl and accounts for over 50% of its total energy. Human milk, on the other hand, is characterized by a low protein content (1.1–1.4 g/dl) and a lactose content around 6.8 g/100 ml. The contribution of lactose to the total milk energy is, however, lower in human milk than in the milk of species such as the donkey and the horse. Although the lactose concentration in the milk of the horse and the donkey is somewhat lower than in human milk, the contribution of lactose to the total milk energy is proportionally greater, due to the low fat content of the milk of these two species.

From the data of human milk composition it follows that the breast-fed human infant receives a relatively high-fat diet. This is in striking contrast with the prenatal situation when most of the energy of the fetus is derived from carbohydrates (60–80%) and proteins, or rather amino acids (20%), with little participation of fat in the oxidative metabolism [2]. Birth, therefore, is associated with the change from a predominantly carbohydrate fuel supply to a high-fat diet.

The limited contribution of protein to the energy value of human milk has been related to the slow growth rate of the human infant, as compared with that of other mammals. *Bunge* [5] is credited to have been the first to note an inverse relationship between the protein content of the milk and the time needed for doubling birth weight in various species. Using more recent data from 14 species I have found an inverse relationship between $\log_{10}$ of milk protein concentration (g/dl) and $\log_{10}$ of birth weight doubling time in days. The correlation coefficient for 29 sets of data was $r = -0.93$. The regression equation derived from these data was: $y = 2.2711 - 1.3681\ x$, where $y = \log_{10}$ birth weight doubling time in days, and $x = \log_{10}$ milk protein concentration in grams per deciliter.

Fat and protein content appear to change in parallel, when milks from different species are compared. However, there seems to be a clear-cut break at about 10–13% protein. The milks of species with such a high protein content have a fat content of the order of 15–20%. However, milks with much higher fat contents, such as those previously mentioned, do not have protein contents higher than about 13 g/dl. It appears, therefore, that the mammary gland is unable to produce milk of a higher protein concentration than about 13%, but it is able to increase further the concentration of fat in the milk, without increasing the concentration of protein.

The relatively high fat content of human milk is believed to be determined by the need of providing a nutriment of a caloric density adequate to the energy requirements of the growing infant in a limited volume. *Fomon* [17] has shown that the infant can adapt the volume consumed to the caloric density of formulas fed *ad libitum*. Infants fed dilute feedings will consume relatively large volumes, whereas infants fed energy-rich feedings will consume relatively small volumes. The ability to make this adjustment depends on the caloric density of the formula and the age of the infant. The ability to adjust the volume of the intake seems to be considerably greater after 41 days of age. However, as *Fomon* pointed out, it is apparent that feedings may be so dilute that even after 112 days of age, the infant is not able to achieve intakes equal to those of infants fed formulas with an energy value of 67 kcal/dl, that is about that of the average human milk.

Fat, because of its high energy values (9.0 kcal or 37.7 kJ/g), has a unique role as a source of energy in infant feeding. Infant formulas with low fat contents, will demand an increase in the volume of the feed which young infants may not be able to achieve. On the other hand, increasing the proportion of protein in the formula would lead to increased solute load on the kidney, while increasing the proportion of carbohydrate could lead to digestive disturbances due to excessive demands on the activity of the disaccharidases of the intestinal mucosa.

It is generally agreed that 850–1,000 ml of good quality human milk per day, maintain satisfactory growth rates in healthy infants, up to the age of 5–6 months [15]. Assuming that such a milk has a fat content of 4.0–4.5 g/dl, we can estimate the daily fat intake between a minimum of 34 g and a maximum of 45 g/day.

Young children fed diets of low energy value often have difficulties in ingesting the large volume of food required to meet their energy needs. This may be solved by increasing the fat content of the diet and, consequently, its energy density. There is, however, an upper limit for the amount of fat in the diet. On the one hand inclusion of large amounts of fat in the diet causes a reduction in the amounts of other dietary items, which may be important as sources of minerals and vitamins. On the other hand, there is evidence that diets deriving 66% or more of their energy from fat, tend to produce ketosis in children who are beyond the rapid growth phase (about 1 year of age). Moreover, it should be kept in mind that children consuming diets which are unsatisfactory both in protein and energy contents will be at risk of not having their protein needs met, if dietary energy is increased only at the expense of fat [6, 15].

Fat Digestion and Absorption in Children

The ability of the gastrointestinal tract to digest and absorb fat is a decisive factor in determining the upper limit of adequate fat content in the infantile diet. This consideration is particularly important in young infants and in prematures. Normal infants may have difficulties with fat absorption during the first month of life [14]. These difficulties are related both to the digestive ability of the infant, which in turn is related to the subject's age, and to the characteristics of the fat. Three factors are important in this regard: (1) the degree of saturation; (2) the molecular size of the fatty acids, and (3) the triglyceride structure.

Influence of Saturation

Generally speaking, fats rich in unsaturated fatty acids (mono and polyunsaturated) are better absorbed than saturated fats. This was first demonstrated by *Tidwell et al.* [57] in 1935 and has been repeatedly confirmed by others. All the authors I have had the opportunity to consult, agree in noting the excellent absorption of unsaturated vegetable oils by children [1, 4, 13, 25, 33, 34, 40, 48, 54, 55]. Their results are in agreement with the classic studies by *Langworthy* [38], who, in 1923, proposed that the absorption of fats depends on their melting points. Vegetable fats, particularly those rich in polyunsaturated fatty acids, have a low melting point and are better absorbed than saturated fats with higher melting points, both by children and by adults, as well as by experimental animals.

Influence of Chain Length

For an equal degree of saturation, short-chain fatty acids tend to be better absorbed than the fatty acids of greater molecular size. With regard to saturated fatty acids, there are numerous observations in the pediatric literature showing that medium-chain fatty acids (8–12 carbon atoms) are easily absorbed. The preparation usually called MCT (medium-chain triglycerides), obtained by fractionation of coconut oil, mainly contains saturated fatty acids, C 8:0 and C 10:0, and it is widely used in the feeding of prematures. It is well absorbed, and when mixed with other fats appears to improve their absorption [56, 64].

Important data in this regard have been recently published by *Bergmann et al.* [3] from the Pediatric Clinic of the Frankfurt University. The results of these authors are in agreement with observations made by pediatricians in the USA and other countries.

Coconut oil, which contains saturated fatty acids between C 8:0 and C 14:0 (mainly C 12:0 and C 14:0), is very well absorbed in spite of its high degree of saturation. Thus *Valyasevi* and his coworkers [personal commun.] in the Departments of Pediatrics and Nutrition of Bangkok University (Thailand), have shown that the absorption of this oil by children 2–3 years of age ranges about 95.8%, which is higher than the absorption of cow's milk fat by infants.

Influence of Triglyceride Structure

A significant progress in our knowledge of fat absorption has been the demonstration that, besides the factors already mentioned, the absorption of fat is determined by the position of the fatty acids in the triglyceride molecule. This finding is important to explain why human milk fat is better absorbed than cow's milk fat. The difference between these two fats with regard to their absorption, has been known for many years [32, 60].

In 1959 *Mattson* [44] compared the absorption of mixtures of tristearin and safflower oil. The mixtures were administered in two different ways: as the simple mixture of the two fats and after the mixture had been submitted to the randomization process. The results showed that the absorption of these two mixtures having identical fatty acid composition, but different distribution of the fatty acids in the triglyceride molecule, was determined by the proportion of triglycerides containing only saturated fatty acids.

Mattson et al. [45] had previously demonstrated that pancreatic lipase selectively splits the fatty acids in positions 1 and 3, but not those in position 2. The triglycerides are absorbed in the form of fatty acids and 2-monoglycerides [45–47]. *Mattson and Volpenhein* [46] also showed that palmitic acid given as the free acid is absorbed only to the extent of 55%. The absorption is 84% when palmitic acid is esterified in position 1, and 94% when esterified in position 2. Thus, monoglycerides are better absorbed than the free fatty acid, and the 2-monoglyceride is better absorbed than the 1-monoglyceride.

The influence of triglyceride configuration on the absorption of fat by the human newborn was studied by *Filer et al.* in 1969 [16]. The absorption of lard was studied in babies 5–9 days of age. The lard was administered in the natural form and after randomization. Structural analysis demonstrated that in natural lard 85.3% of the palmitic acid is esterified in position 2 of the glyceride, whereas in the randomized lard only 33% of palmitic acid is esterified in that position. The feeding experiments demonstrated that natural lard was absorbed to the extent of 95%, but only 72% of the randomized lard was absorbed. Calculation of the absorption of individual fatty acids demonstrated that palmitic acid was absorbed to the extent of 92% when given as natural lard, and only to the extent of 58% when given as randomized lard. Thus, the absorption of palmitic acid is clearly dependent on the position of the acid in the triglyceride molecule. Linoleic acid, on the the other hand, was absorbed to the extent of 98% when given as natural lard, and to 91% when given as randomized lard.

These results, as already mentioned, are important in explaining the absorption differences between the fats of human and bovine milks. As shown by *Mc Carthy et al.* [41] and by *Freeman et al.* [20] palmitic acid is uniformly distributed among the three positions in cow's milk fat, whereas most of the palmitic acid is esterified in position 2 in the fat of human milk.

The better absorption of human milk fat, as compared to the fat of cow's milk, can be explained, at least in part, by the preferential esterification of palmitic acid in position 2 in the fat of the human milk.

The influence of the position of the saturated fatty acids in the triglyceride molecule on fat absorption is not limited to the human species. Studies reported by *Tomarelli et al.* [59] indicate rhat rats absorb 94.6% of human milk fat, but only 89.5% of cow's milk fat. The absorption of palmitic acid showed a direct lineal relationship with the proportion of palmitic acid esterified in position 2. The proportions of palmitic acid esterified in position 2, found by *Tomarelli et al.* [59] were, 84% for lard, 68% for human milk fat, and 43% for cow's milk fat.

Difficulties with fat absorption are particularly important in premature babies. These babies are able to absorb human milk fat to the extent of about 80% [3], but their ability to absorb cow's milk fat is smaller. Thus, in the review I have made of 19 studies including a total of 185 prematures, the maximal absorption of cow's milk fat was 82.9% and the lowest 33.4%. More than half of the absorption values reported were between 50 and 70%, which seems to indicate that the average absorption of cow's milk fat by prematures must be about 60%. Given the heterogeneous nature of the data, no statistical analysis of the results was attempted.

Lipolytic Enzymes in Milk

Because digestion and absorption of lipids is less efficient in newborns than in adults [14], the presence of lipolytic enzymes in milk could be an important contribution to the digestion of dietary lipids, in infants with low endogenous lipase activity.

It has been shown [28, 29, 51] that human milk contains an enzyne which, in the presence of certain bile salts, has a high lipase and esterase activity. The enzyme has been found only in milk from primates; it is inactivated by pasteurization of the milk, and its lipolytic activity is inhibited by eserine. In a recent paper, *Fredrikzon et al.* [19] have shown in infants fed fresh human milk, that a fraction of the lipase activity of the duodenal contents was inhibited by eserine. The degree of inhibition corresponded to the lipase activity of the milk fed. The duodenal content collected 1 h after feeding fresh human milk, was found to hydrolyze retinol palmitate severalfold faster than the duodenal content obtained after feeding pasteurized human milk.

The Swedish suthors [19] conclude that the bile-salt-stimulated lipase from human milk is active in the intestine, and that it may contribute to the digestion of the milk lipids, particularly the retinol esters. Since the retinol esters in milk are the source of vitamin A for the newborn, and must be hydrolyzed before they are absorbed, the bilesalt-stimulated lipase from human milk may have an important role in enhancing vitamin A absorption in the infant. This may be particularly significant for babies nursed by malnourished mothers secreting milk with low vitamin A content.

Essential Fatty Acid Requirements in Childhood

Fatty acid analyses of milk have consistently shown that the linoleic acid content of human milk fat is significantly higher than that of cow's milk. Figures commonly quoted [17] indicate that the linoleic acid (C 18:2n6) of cow's milk fat (as percent of the total fatty acids) is of the order of 2%, whereas that of human milk fat is close to 10.0% [7]. Analyses currently performed in our laboratory indicate that the average linoleic acid content of the milk fat from mothers of this part of Spain, consuming their usual diets, approaches 15% of the total fatty acids. Since *Potter and Nestel* [52] have clearly shown the influence of dietary linoleic acid on the linoleic acid content of human milk, these results may reflect a high intake of linoleic acid by these mothers.

Crawford et al. [10] have noted that the pattern of polyunsaturated fatty acids of human milks is close to that of human brain lipids. The essential fatty acids (EFA) of the brain in the mature human fetus are mainly C 20:4n6 and C 22:4n6. These fatty acids can be formed in the body from linoleic acid, which, however, appears in a low proportion in the brain lipids.

α-Linolenic acid (C 18:3n3) is considered as the precursor of the n-3 long-chain polyunsaturated fatty acids found in brain ethanolamine phosphoglycerides, and in the lipids of the retina [8, 9].

In their recent review on the EFA requirements in infancy, *Crawford et al.* [9] point out that the long-chain (C20–C22) polyunsaturated fatty acids would contribute greatly to the EFA potency of milk. Their higher level in the human milk would therefore contribute to its higher EFA activity, as compared to that of cow's milk. As noted by *Crawford et al.* [9], this fact has been neglected by *Cuthbertson* [11] in this review on EFA requirements in infancy.

The EFA requirements of the human newborn have been recognized for the last 15 years or so. In 1963, *Hansen et al.* [27] demonstrated the development of clinical signs of EFA deficiency in children fed diets devoid of

Table I. Recommended intakes for infants; as percent of total energy intake

Reference	EFA intake
Combes et al., 66	4.5
Holman et al., 67	1.4
Paulsrud et al., 69	> 2.0
Houtsmüller, 68	5.0
American Academy of Pediatrics, 65	3.0
Codex Alimentarius, 6	2.7
FAO/WHO Expert Consultation, 15	3.0

EFA. These signs were corrected by the administration of trilinolein. More recently, *Friedman et al.* [21] have clearly shown the indispensability of EFA in children receiving intravenous feeding. *Holman* [31] has concluded that: 'The prolonged use of the usual fat-free intravenous preparations will surely precipitate an EFA deficiency, creating new and unknown consequences and hazards, causing a nutritional deficiency while attempting to avoid others.'

Some recommendations regarding EFA intakes for infants are summarized in table I.

With the data currently available, it is easy to see that the EFA content of human milk (including its content of long-chain polyunsaturated fatty acids) is generally well above these recommendations. Cow's milk, however, seems to be far below these recommendations, with an EFA content usally estimated to correspond to no more than 1% of its total energy. It is therefore pertinent to ask whether infants exclusively fed with cow's milk are in danger of developing EFA deficiency. *Naismith et al.* [50] have compared 11 infants fed exclusively on a cow's milk formula and 5 wholly breast-fed infants for 14 weeks. They conclude from this comparison that: 'Infants fed solely on cow's milk formula over a realistic period of time before weaning, are not at risk from linoleic acid deficiency.'

The data given by these authors, however, show that the linoleic acid content of the total plasma lipids in the breast-fed children was about 166% of that observed in those fed cow's milk. Arachidonic acid (C 20:4n6) was about twice in the breast-fed children, as compared with those fed on cow's milk. Data on the long-chain polyunsaturated fatty acid content of the erythrocyte lipids from these children have been reported by *Sanders and Naismith* [53]. All these data show that the EFA levels in the blood lipids of infants fed on cow's milk preparations are lower than those observed in the breast-fed infants. Furthermore, the levels of 5, 8, 11-eicosatrienoic acid (C 20:3n9), a nonessential fatty acid produced in EFA deficiency states, are elevated. It is then clear that although bottle-fed babies did not show clinical manifestations of EFA deficiency, the fatty acid composition of their blood lipids was different from that observed in the breast-fed babies, showing changes in fatty acid distribution which are in the same direction as those observed in EFA deficiency [30]. The data reported by *Crawford et al.* [8] indicate that the triene/tetraene ratio (that is the ratio of C 20:3n9 to C 20:4n6) in the blood phospholipids is below 0.05 in breast-fed babies, but reaches levels of 0.2–0.3 in bottle-fed babies.

In their recent review on EFA requirements in infancy, [9] *Crawford et al.* [9] conclude that it is rather difficult to exclude EFA deficiency from the factors involved in the health risks known to affect the bottle-fed infant.

In view of these data, the recommendation of a dietary intake of EFA, equivalent to at least 3% of the total energy intake, seems to be justified. A proportion of EFA similar to that found in human milk is currently recommended for milk preparations intended as substitutes for human milk in infant feeding [6, 15].

Dietary Fat in the Control of Plasma Lipid Levels

Because of the statistical association between plasma cholesterol level and the risk of developing coronary heart disease, and because the plasma cholesterol levels are influenced by the amount and fatty acid composition of the fat in the diet, there is much current interest in the role of fat in the control of plasma lipid levels in man.

Most of the information regarding the effect of dietary fat on plasma cholesterol concentration has been obtained in adults, and there is general agreement that the glycerides of saturated fatty acids increase plasma cholesterol, whereas polyunsaturated fatty acid glycerides have the opposite effect. Monounsaturated fatty acid glycerides are neutral in this respect, that is, comparable to the mixed carbohydrates of the diet taken as reference [22, 36, 37, 42]. Regarding the cholesterol-depressing effect of polyunsaturated fatty acids in man, no difference was found between the effect of linoleic acid and that of the polyunsaturated fatty acids of fish oil [26].

There is less quantitative information regarding the effect of dietary fat on the plasma lipid levels in infancy and childhood, but there is good evidence that hypercholesterolemic infants fed a formula containing a vegetable oil rich in linoleic acid, have lower plasma cholesterol levels at 6 months of age than children of the same ages fed a formula containing cow's milk fat [23]. At 16 weeks of age, normolipidemic children fed the cow's milk formulas usual in the USA have plasma cholesterol levels 40 mg/dl higher than children fed formulas based on skim milk and corn oil [39]. The data contained in some of the reports [12, 63] indicate that breast-fed babies have higher plasma cholesterol than those fed cow's milk, in spite of the higher linoleic acid content of the former. However, *Potter and Nestel* [52] have shown an inverse relationship between the plasma cholesterol level of the baby and the linoleic content of the mother's milk.

Several reports indicate that plasma cholesterol levels in school children and adolescents are related to the composition of the diet [24, 61]. *McGandy et al.* [43] and *Ford et al.* [18], working with boys living in boarding schools have shown that reducing saturated fatty acid glycerides and cholesterol and increasing polyunsaturated fatty acid glycerides in the school diet, led to a prompt decrease in mean plasma cholesterol levels, of 10% in one school and of 15% in another.

In a similar experiment with institutionalized boys and girls, *Vergroesen and de Boer* [62], showed substantial decreases of plasma cholesterol, when the fats of the usual institutional diet were replaced by fat products containing large amounts of linoleic acid. The reduction in plasma cholesterol was from 192 to 149 mg/dl in the first part of the experiment, and from 198 to 160 mg/dl in the second part.

As in the experiments of *McGandy et al.* [43] and *Ford et al.* [18], no complete dietary adherence was assured. I believe, therefore, that *McGandy and Hegsted* [42], are justified in concluding that these are probably not the maximal responses that can be achieved by a similar dietary modification.

It appears, therefore, that plasma cholesterol levels can be reduced by modifying the fatty acid composition of the diet in infants and school children. This role of fat is of obvious interest in contemporary child nutrition in view of the emphasis currently laid on the early prevention of atherosclerosis [49]. A complete analysis of the problem of dietary prevention of atherosclerosis, starting in childhood is, however, beyond the scope of this discussion.

References

1   Barltrop, O. and Oppe, Th.E.: Absorption of fat and calcium by low-birthweight infants from milks containing butterfat and olive oil. Archs. Dis. Childh. *48:* 496 (1973).

2   Battaglia, F.C. and Meschia, G.: Principal substrates of fetal metabolism. Physiol. Rev. *58:* 499 (1978).

3   Bergmann, R.L.; Jäger, H.; Kern, M. und Bergmann, K.E.: Mittelkettige Triglyceride in der Ernährung von Frühgeborenen. Mschr. Kinderheilk. *125:* 953 (1977).

4   Breslow, L.: A clinical approach to infantile colic. J. Pediat. *50:* 196 (1957).

5   Bunge, G.V.: Physiologie des Menschen, vol II, p. 119 (Vogel, Leipzig, 1901).

6   Codex Alimentarius Commission. Recommended international standards for foods for infants and children. Joint FAO/WHO Food Standards programme FAO, Rome WHO, Geneva (1976).

7   Crawford, M.A.; Hall, B.; Laurance, B.M., and Munhambo, A.: Milk lipids and their variability. Curr. med. Res. Opin. *4:* suppl. 1, p. 33 (1976).

8   Crawford, M.A.; Hassam, A.G., and Hall, B.M.: Metabolism of essential fatty acids in the human fetus and neonate. Nutr. Metab. *21:* suppl. 1, p. 187 (1977).

9   Crawford, M.A.; Hassam, A.G., and Rivers, J.P.W.: Essential fatty acid requirements in infancy. Am. J. clin. Nutr. *31:* 2181 (1978).

10 Crawford, M.A.; Sinclair, A.J.; Msuya, P.M., and Munhambo, A.: Structural lipids and their polyenoic constituents in human milk; in Galli, Jacini and Pecile, Dietary lipids and postnatal development, p. 41 (Raven Press, New York 1973).

11 Cuthbertson, W.F.J.: Essential fatty acid requirements in infancy. Am. J. clin. Nutr. *29:* 559 (1976).

12 Darmady, J.M.; Fosbrooke, A.S., and Lloyd, J.K.: Prospective study of serum cholesterol levels during the first year of life. Br. med. J. *ii:* 685 (1972).

13 Davidson, M. and Bauer, C.H.: Patterns of fat excretion in feces of premature infants fed various preparations of milk. Pediatrics, Springfield *25:* 375 (1960).

14 Espgan Committee on Nutrition: Guidelines on infant nutrition. Acta paediat. scand., suppl. 262 (1977).

15 FAO Report on expert consultation. Dietary fats and oils in human nutrition FAO, Rome (1978).

16 Filer, L.J.; Mattson, F.H., and Fomon, S.J.: Triglyceride configuration and fat absorption by the human infant. J. Nutr. *99:* 293 (1969).

17 Fomon, S.J.: Infant nutrition; 2nd ed. (Saunders, Philadelphia 1974).

18 Ford, C.H.; Mc Gandy, R.B., and Stare, F.J.: An institutional approach to the dietary regulation of blood cholesterol in adolescent males. Prev. Med. *1:* 426 (1972).

19 Fredrikzon, B.; Hernell, O.; Blackberg, L., and Olivecrona, T.: Bile salt-stimulated lipase in human milk: evidence of activity in vivo and of a role in the digestion of milk retinol esters. Pediat. Res. *12:* 1048 (1978).

20 Freeman, C.P.; Jack, E.L., and Smith, L.M.: Intramolecular fatty acid distribution in the milk fat triglycerides of several species. J. Dairy Sci. *48:* 853 (1965).

21 Friedman, Z.; Danon, A.; Stahlman, M.T., and Oaten, J.A.: Rapid onset of essential fatty acid deficiencies in the newborn. Pediatrics, Springfield *58:* 640 (1976).

22 Glueck, C.J. and Connor, W.E.: Diet-coronary heart disease relationships reconnoitered. Am. J. clin. Nutr. *31:* 727 (1978).

23 Glueck, C.J. and Tsang, R.C.: Pediatric type II hyperlipoproteinemia. Effects of diet on plasma cholesterol in the first year of life. Am. J. clin. Nutr. *25:* 224 (1972).

24 Golubjatnikov, R.; Paskey, T., and Inborn, S.L.: Serum cholesterol levels in Mexican and Wisconsin school children. Am. J. Epidemiol. *96:* 36 (1972).

25 Gordon, H.H. and Mc Namara, H.: Fat excretion in premature infants. I. Effects on fecal fat of decreasing fat intake. Am. J. Dis. Child. *62:* 328 (1941).

26 Grande, F.; Anderson, J.T., and Keys, A.: Comparison of effects of fish oil polyunsaturated fatty acids and of linoleic acid on man's serum lipids. Am. J. clin. Nutr. *12:* 331 (1963).

27 Hansen, A.D.; Wiese, H.F.; Boelsche, A.N.; Haggard, M.E.; Adam, D.J.D., and Davis, H.: Pediatrics, Springfield *31:* suppl. 1, p. 171 (1963).

28 Hernell, O.: Human milk lipases. III. Physiological implications of bile salt-stimulated lipase. Eur. J. clin. Invest. *5:* 267 (1975).

29 Hernell, O. and Olivecrona, T.: Human milk lipases. II. Bile salt-stimulated lipase. Biochim. biophys. Acta *369:* 234 (1974).

3o Holman, R.T.: The ratio of trienoic:tetraenoic acids in tissue lipids as a measure of essential fatty acid requirements. J. Nutr. *70:* 405 (1960).

31 Holman, R.T.: Essential fatty acid deficiency in humans; in Galli, Jacini and Pecile, Dietary lipids and postnatal development, p. 127 (Raven Press, New York (1973).

32 Holt, L.E.; Courtney, E.M., and Fales, H.L.: Fat metabolism of infants and young children. II. Fat in the stools of infants fed on modifications of cow's milk. Am. J. Dis. Child. *17:* 423 (1919).

33 Jappich, G.; Löhr, H. und Wolf, H.: Fettbilanzstudien mit fettausgetauschter Milch. Z. Kinderheilk. *82:* 7 (1959).

34 Karte, H. und Wolf, H.: Vergleichende Ernährungsversuche und Bilanzuntersuchungen bei Frühgeborenen; in Willi, Symp. Ernährung der Frühgeborenen, Bad Schachen, 1964, p. 91 (Karger, Basel 1965).

35 Kayden, H.J.; Senior, J.R., and Mattson, F.H.: The monoglyceride pathway of fat absorption in man. J. clin. Invest. *46:* 1695 (1967).

36 Keys, A.; Anderson, J.T., and Grande, F.: Serum cholesterol responses to diet. Metabolism *14:* 747, 759, 766 and 776 (1965).

37 Keys, A.; Grande, F., and Anderson, J.T.: Bias and misrepresentation revisited: perspective on saturated fat. Am .J. clin. Nutr. *27:* 188 (1974).

38 Langworthy, C.F.: The digestibility of fats. Ind. Engng. Chem. *15:* 276 (1923).

39 Lowe, C.J.; Mosovich, L.L., and Pessin, V.: Effect of protein level and type of milk formulas on growth and maturation of infants. J. Pediat. *64:* 666 (1964).

40 Luther, G. und Schreier, K. :Untersuchungen zur Resorption einzelner Fettsäuren an Säuglingen. Klin. Wschr. 41: 189 (1963).

41 Mc Carthy, R.D.; Patton, S., and Evant, L.: Structure and synthesis of milk fat. II. Fatty acid distribution in the triglycerides of milk and other animal fats. J. Dairy Sci. *43:* 1196 (1960).

42 Mc Gandy, R.B. and Hegsted, D.M.: Quantitative effects of dietary fat, and cholesterol on serum cholesterol in man; in Vergroesen, The role of fats in human nutrition, p. 211. (Academic Press, London 1975).

43 Mc Gandy, R.B.; Hall, B.; Ford, C., and Stare, F.J.: Dietary regulation of blood cholesterol in adolescent males.: A pilot study. Am J. clin. Nutr. *25:* 61 (1972).

44 Mattson, F.H.: The absorbability of stearic acid when fed as a simple or mixed triglyceride. J. Nutr. *69:* 338 (1959).

45 Mattson, F.H.; Benedict, J.H.; Martin, J.B., and Beck, L.W.: Intermediates formed during the digestion of triglycerides. J. Nutr. *48:* 335 (1952).

46 Mattson, F.H. and Volpenhein, R.A.: Rearrangement of glyceride fatty acids during digestion and absorption. J. biol. Chem. *237:* 53 (1962).

47 Mattson, F.H. and Volpenhein, R.A.: The digestion and absorption of triglycerides. J. biol. Chem. *239:* 2772 (1964).

48 Milner, R.D.G.; Deodehar, Y.; Chard, C.R., and Grant, R.M.: Fat absorption by small babies fed two filled milk formulas. Archs Dis. Childh. *50:* 654 (1975).

49 Mitchell, S.: Symposium on prevention of atherosclerosis at the pediatric level. Am. J. Cardiol. *31:* 539 (1973).

50 Naismith, D.J.; Deeprose, S.P., and Ma, M.C.F.: The linoleic acid requirements of the human infant. Proc. Nutr. Soc. *35:* 65A (1976).

51 Olivecrona, T. and Hernell, O.: Human milk lipases and their possible role in fat digestion. Paediat. Paedol. *11:* 600 (1976).

52 Potter, J.M. and Nestel, P.J.: The effect of dietary fatty acids and cholesterol on the milk lipids of lactating women and the plasma cholesterol of breast-fed infants. Am. J. clin. Nutr. *29:* 54 (1976).

53 Sanders, T.A.B. and Naishmith, D.J.: Long-chain polyunsaturated fatty acids in the erythrocyte lipids of breastfed and bottle-fed infants. Proc. Nutr. Soc. *35:* 63 A (1976).

54 Schreier, K.: Fettstoffwechselstudien im frühen Kindesalter. Über die Resorption von Fetten mit einem hohen Gehalt an ungesättigten Fettsäuren. Z. Kinderheilk. *81:* 442 (1958).

55 Shaw, J.C.L.: Evidence for defective skeletal mineralization in low birth weight infants: the absorption of calcium and fat. Pediatrics, Springfield, *57:* 16 (1976).

56 Tantibhedhyangkul, P. and Hashim, S.A.: Medium-chain triglyceride feeding in premature infants: effects on fat and nitrogen absorption. Pediatrics, Springfield *55:* 539 (1975).

57 Tidwell, H.C.; Holt, E.L.; Farrow, H.L., and Neals, S.: Studies in fat metabolism. II. Fat absorption in premature infants and twins. J. Pediat. *6:* 481 (1935).

58 Tiscornia, E.: Attuali conoscenza sulla composizione chimica della latte alimentare. (I–III). Riv. Soc. ital. Sci. Aliment. *6:* 25, 291, 423 (1977).

59 Tomarelli, R.M.; Meyer, B.J.; Weaber, J.R., and Beruhart, F.W.: Effect of positional distribution on the absorption of the fatty acids of human milk and infant formula. J. Nutr. *95:* 583 (1968).

60 Van de Kamer, J.H. and Weijers, H.A.: Malabsorption syndrome. Fed. Proc. *20:* 335 (1961).

61 Van der Haar, F. and Kromhout, D.: Food intake, nutritional anthropometry and blood chemical parameters in 3 selected Dutch schoolchildren populations. Mededeling Landbouwhogeschool Wageningen, Nederland, 78–9 (1978) (Veenman & Zonen, Wageningen 1978).

62 Vergroesen, A.J. und De Boer, J.: Quantitative und qualitative Effekte mehrfach ungesättigter und anderer Fettsäuren in der menschlichen Diät; in Polyenfettsäuren, vol. 22, p. 76 (Steinkopff, Darmstadt 1971).

63 Woodruff, C.W.; Bailey, M.C.; Davis, J.T., et al.: Serum lipids in breast-fed infants and in infants fed evaporated milk. Am. J. clin. Nutr. *14:* 83 (1964).

64 Yamashita, F., Shibuya, S. and Funatsu, I.: Absorption of medium chain triglyceride in the low birth weight infant and evaluation of MCT milk formula for low weight infant nutrition using latin square technique. Kurume med. J. *16:* 191 (1969).

65 American Academy of Pediatrics, Committee on Nutrition: 1976 Recommendations. Nutr. Rev. *34:* 248 (1976).
66 Combes, M.A.; Pratt, E.L., and Wiese, H.F.: Essential fatty acids in premature infant feeding. Pediatrics, Springfield *30:* 136 (1962).
67 Holman, R.T.; Caster, W.O., and Wiese, H.F.: The essential fatty acid requirement of infants and the assessment of their dietary intake of linoleate by serum fatty acid analysis. Am. J. clin. Nutr. *14:* 70 (1964).
68 Houtsmüller, U.M.T.: Evaluation of modern foods as sources of lipids; in Lipids, malnutrition and the developing brain. Ciba Fdn Symp., p. 213 (North Holland/ Elsevier/Excerpta Medica, Amsterdam 1972).
69 Paulsrud, J.R.; Pensler, L.; Whitten, C.F.; Srewart, S., and Holman, R.T.: Essential fatty acid deficiency in infants induced by fat-free intravenous feeding. Am. J. clin. Nutr, *25:* 897 (1972).

Prof. F. Grande, Institute of Biochemistry and Nutrition, Fundación
F. Cuenca Villoro, Gascon de Gotor 4, Zaragoza (Spain)

Discussion

Prof. Verdonk: I would like to add that future mothers who have received a diet enriched in linoleic acid not only increased their milk concentration in polyunsaturated fatty acids but also increased their total milk secretion.

Prof. Zöllner: If you analyse the blood from the cord at term, you have the biochemical picture of latent essential fatty acid deficiency. Is transport across the placenta membrane not sufficient to produce normal linoleic acid levels in the blood, levels which are rapidly obtained when the child is breast-fed?

Dr. Crawford: Low levels of linoleic acid in cord blood are due to the fact that the fetus, and possibly the placenta, desaturate and elongates the chain of linoleic acid to arachidonic acid. This means that the linolelc acld content in cord blood is lowei than in maternal blood.

However, if you analyse the arachidonic acid, you find it is much higher in cord blood. The same applies to the relationship between linolenic and docosahexaenoic acids. The low levels of linoleic acid in cord blood do not mean that the fetus is deficient in essential fatty acid. This is, in our view, evidence of the fetal conversion of linoleic and linolenic acids to arachidonic and docosahexaenoic acids which are specifically used for the developing brain. The brain uses arachidonic and docosahexaenoic acids and not linoleic and linolenic acids. In the human, some 70% total adult value of brain cells divide during fetal growth. The conversion of linoleic and linolenic acids to their long-chain derivates provides the appropriate fatty acids for the developing brain cell membranes which is, of course, of major consideration of human fetal growth.

Nutr. Metab. *24* (Suppl. 1) 162–180 (1980)

Cardiovascular Effects of Dietary Linoleic Acid

F. ten Hoor

Unilever Research, Vlaardingen

Key Words. Dietary linoleic acid · Prostaglandins · Arterial thrombosis · Blood pressure · Heart function

Abstract. A survey is given of the effects of a linoleic-acid-rich diet on arterial thrombosis, blood platelet function and blood pressure in rats and man, and on coronary flow rate and heart muscle function of isolated rat hearts. Dietary linoleic acid decreases arterial thrombosis tendency in a dose-dependent manner in rats, as measured by the aorta loop technique. In a group of patients, a linoleic-acid-rich diet improved blood platelet aggregation measured with the filtragometer as compared with a similar group which was given a linoleic-acid-poor diet. The patients who received a high linoleic acid diet showed a lower cardiovascular death rate than the low linoleic acid group. Salt-induced hypertension in rats can be prevented and cured by a high linoleic acid diet. In man, the results of a few experiments show a decreasing tendency of blood pressure during a diet enriched with linoleic acid. In the isolated hearts of rats fed a diet high in linoleic acid, coronary flow and heart muscle function both increase as compared to control groups fed lard or hardened coconut oil. The observed effects are discussed against the background of recent findings in the field of prostaglandin research.

Since the classical work of *Burr and Burr* [1], linoleic acid has been known as an essential fatty acid which must be ingested with food because it cannot be synthesized by the animal organism. As for other essential food components, e.g. essential amino acids, vitamins and minerals, a minimum requirement has also been determined for linoleic acid. Based on either biochemical [2] or biological [3, 4] criteria, this requirement was estimated to be about 2% of the total energy intake [for a review, see *Crawford et al.,* 5]. However, results of investigations in the field of atherosclerosis revealed that a much larger amount of linoleic acid than the recommended minimum is necessary to lower a high blood cholesterol concentration, one of the main risk factors in the development of atherosclerosis [for a review, see *Jackson et al.,* 6]. It was thus shown that the minimum requirement, as determined on the basis of several biochemical and biological criteria is far from an optimum under other circumstances.

During the last 15 years, a number of investigators systematically studied whether dietary linoleic acid in amounts exceeding the so-called minimum requirement has physiological effects on organs and organ systems, the dysfunction of which is closely related to the development of cardiovascular diseases. It is the purpose of this review to give a survey of these effects on arterial thrombosis and blood platelet function, blood pressure, coronary flow and heart muscle function and on the function of the arterial vessel wall. The results presented will be discussed with respect to recent developments in the field of prostaglandin research.

Arterial Thrombosis and Blood Platelet Function

Using the aorta loop technique [7] *Hornstra* [8] investigated the influence of various dietary fats on the tendency to arterial thrombosis in rats. The results of these experiments can be summarized as follows: dietary fats containing mainly saturated fatty acids with more than 12 C atoms enhance arterial thrombus formation; dietary linoleic acid inhibits the tendency to arterial thrombosis in a dose-dependent manner; oleic-acid-like fatty acids are neutral in this respect. Figure 1 shows the method used and illustrates the effects of the various dietary fatty acids in this animal model.

In man, an impression of arterial thrombosis tendency may be obtained by measuring blood platelet function. The property of blood platelets most often used for this is aggregation, which can be determined in several ways. The filtragometer technique [9] measures blood platelet aggregation by drawing blood with a constant flow rate from a cubital vein through a thin metal filter with pores of 20 μm square. The resistance of the filter is determined by the amount of occluded pores and can be monitored by continuously recording the pressure difference over the filter. As the blood is heparinized just after it has left the vein, occlusion of the pores is essentially caused by blood platelet aggregates (fig. 2). The time necessary to reach an arbitrarily defined pressure difference over the filter is the aggregation time which is taken as a measure for the aggregation tendency of blood platelets; a long aggregation time represents a weak, a short aggregation time a strong aggregation tendency.

Figure 3a shows the principle of the method, figure 3b the results of the filtragometer response in a long-term clinical trial in which two groups of

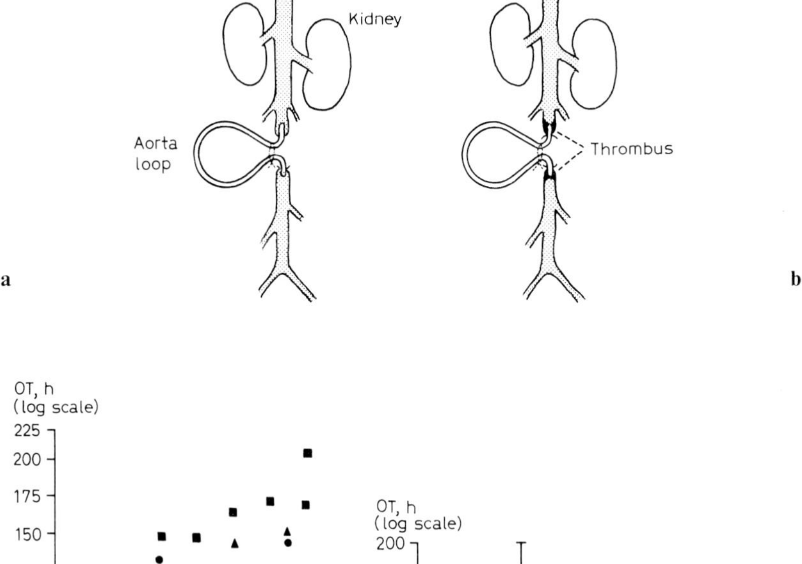

Fig. 1. Effects of dietary fats on aorta-loop-induced arterial thrombosis in rats. **a** A polyethylene cannula is inserted into the abdominal aorta of the rat under general anaesthesia. The loop partially protrudes from the abdominal wall, which is closed after insertion of the loop. **b** After a few days, the circulation through the loop will be prevented due to thrombus formation at the sites of insertion of the loop. Blood flow through the loop coming to a standstill can be observed in the extra abdominal part, and the time between insertion of the loop and its obturation (obturation time, OT) is determined. **c** After feeding 50 en% of various dietary fats for 8 weeks to groups of rats, aorta loops were inserted and the obturation times determined. As shown, a higher intake of linoleic acid (C 18:2) results in a longer obturation time indicating a lower thrombosis tendency. ■ = Sunflowerseed oil; ▲ = mixtures of sunflowerseed oil and hardened coconut oil; ● = mixtures of sunflowerseed oil and butterfat. **d** Four groups of rats were fed a diet containing 60 en% fat. Saturated fatty acids (C 16:0 + C 18:0), linoleic acid (C 18:2) and oleic acid (C 18:1) concentrations were varied as shown (% of total amount of fat). It is clear that feeding the high linoleic acid diets lengthens obturation time regardless of the amount of dietary saturated fatty acids. Oleic acid does not influence the results.

Fig. 2. Scanning electron-microscopic picture of a partly occluded filtragometer filter. Size of pores 20 μm square. × 800.

age- and sex-matched patients were given for 12 years diets containing 4 and 12 en% linoleic acid respectively [10]. The patients receiving 12 en% linoleic acid had on the average a longer aggregation time – and thus a smaller aggregation tendency – than the patients receiving 4 en% linoleic acid [11]. The group of patients who got a high linoleic acid diet showed a lower cardiovascular death rate than the low linoleic acid diet group [10].

It thus appears that dietary fat can influence arterial thrombosis tendency in rats and blood platelet function in man. Dietary linoleic acid apparently improves platelet function resulting in a decrease in thrombotic complications.

Blood Pressure

The blood-pressure-lowering effects of dietary linoleic acid on salt-induced hypertension in rats have been described by *Triebe et al.* [12] and *Ten Hoor and Van de Graaf* [13]. The results of a typical crossover experiment

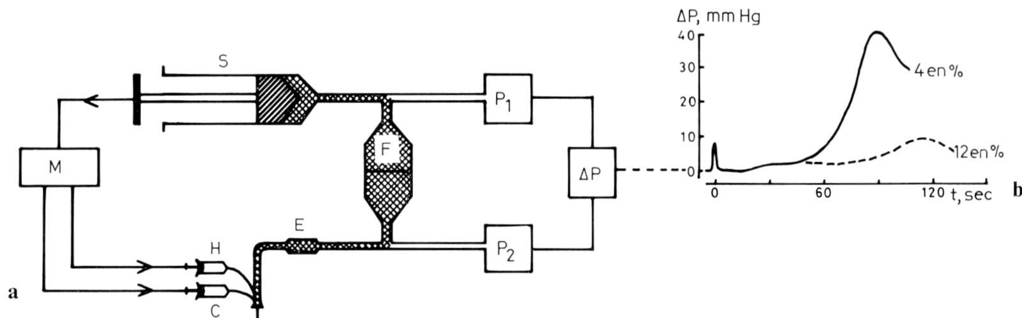

Fig. 3. a Diagram of the filtragometer. Blood from a cubital vein is drawn with a motor (M)-driven syringe (S) via a mixing chamber (E) through a nickel filter (F) with pores of 20. Heparin (H, end concentration 5 IU·ml⁻¹) or citrate (C) prevents clotting of blood. The pressure difference over the filter (ΔP) is continuously recorded. The aggregation time is defined as the time period between the start of the experiment and the moment at which $\Delta P = 5$ mm Hg. **b** Mean filtragometer tracings of two groups of patients, receiving for 12 years diets containing 4 and 12 en% linoleic acid respectively [10, 11]. The aggregation time of the high linoleic acid group (82 patients) is longer (p < 0.001) than that of the low linoleic acid group (62 patients) indicating a smaller blood platelet aggregation tendency.

are presented in figure 4 which shows that a diet containing 35% of total energy (35 en%) as sunflowerseed oil (with 70% linoleic acid) keeps blood pressure low, when rats are loaded with 1.5% NaCl in the drinking water (which means about 1 g NaCl/kg/day!). On the contrary, blood pressure is rising when a diet containing 35 en% coconut oil (with 2% linoleic acid) is given. The crossover shows that the blood pressure effect is clearly diet dependent.

Another striking fact from this experiment is the observation that after stopping NaCl loading, the type of diet no longer seems to have any influence on blood pressure. Figure 5 shows the results of a similar experiment. In addition to dietary effects on salt-loaded hypertension, in this experiment the effect of the prostaglandin synthetase inhibitor acetyl salicylic acid was also investigated, as it is well documented that prostaglandins play an important role in blood pressure regulation [14]. In both dietary groups, during salt loading, acetyl salicylic acid increases blood pressure as compared with their respective control group, but there is no effect of either diet or acetyl salicylic acid after salt loading has been stopped. When looking at the urinary prostaglandin metabolites, the most striking phenomenon is that

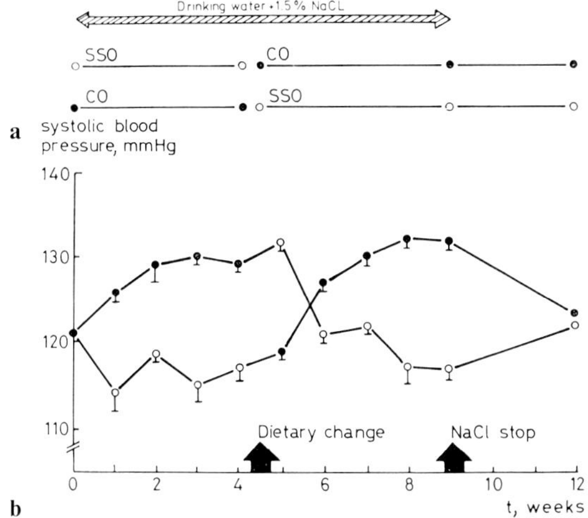

Fig. 4. Influence of dietary fat on NaCl-induced hypertension in rats. **a** Plan of the experiment. NaCl (1.5%) was added to the drinking water for 9 weeks. Two groups of rats (n = 8) with the same mean systolic blood pressure (tail cuff method) were fed diets containing 40 en% sunflowerseed oil (SSO; 65% C 18:2) or 40 en% coconut oil (CO; 2% C 18:2). After 4.5 weeks diets were switched. **b** Course of systolic pressure. It is clear that blood pressure during NaCl loading only rises in the coconut oil group. In the sunflowerseed oil group, blood pressure remains low. The increased blood pressure in the coconut oil group is readily decreased when replacing coconut oil by sunflowerseed oil. The vertical bars indicate the SEM.

the high linoleic acid group has a high urinary excretion of prostaglandin metabolites during the whole experiment. In the low linoleic acid group, the amount of prostaglandin metabolites is gradually decreasing in the course of the experiment. During salt loading, acetyl salicylic acid results in a very low output of prostaglandin metabolites in both dietary groups. However, after salt loading has been stopped, the groups fed the low linoleic acid diet have a low output of prostaglandin metabolites, while the groups fed the high linoleic acid diet show a high output.

Iacono et al. [16, 17] and *Comberg et al.* [18] reported blood-pressure-lowering effects of a high linoleic acid diet in patients with 'borderline' hypertension (about 140/90 mm Hg). The results of these experiments are presented in table I. In 'The National Diet-Heart Study' [19], blood pressures decreased during the experimental period in all the diet groups investigated (P/S ratios ranged from 0.4–4.4). However, the mean initial blood pressure

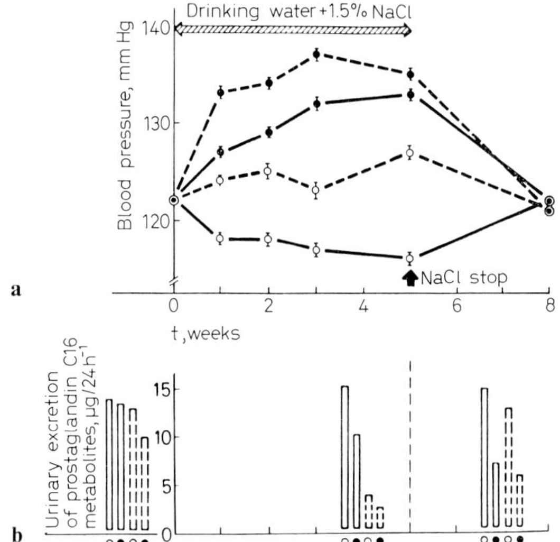

Fig. 5. Influence of a linoleic acid-rich diet and of acetyl salicylic acid on blood pressure after administration of NaCl in the drinking water (a) and on urinary excretion of prostaglandin C16 metabolites in rats (b). Urine was collected for 24 h and was pooled per 4 rats. Drterminations (in duplicate) were done according to a method described by *Nugteren* [15]. Until t = 0 the animals were fed pellets and then divided into four groups (n = 8). O = Sunflowerseed oil, 40 en%; ● = coconut oil, 40 en%; – – – – = acetyl salicylic acid, 30 mg/rat/day. Data on diets and blood pressure measurements as described in figure 4. The vertical bars indicate the SEM.

Table I. Effect of a linoleic acid-enriched diet on arterial blood pressure (mm Hg) in man

Data from *Iacono et al.* [16] n = 21			Data from *Comberg et al.* [18] n = 8		
time	systolic	diastolic	time	systolic	diastolic
initial	136 ± 4	80 ± 3	initial	135 ± 10	92 ± 9
40 days	123 ± 3	73 ± 2	14 days	131 ± 8	88 ± 5
80 days	124 ± 3 diet stopped	76 ± 2	28 days	127 ± 7 diet stopped	85 ± 7
113 days	129 ± 3	83 ± 2	118 days	136 ± 13	91 ± 7

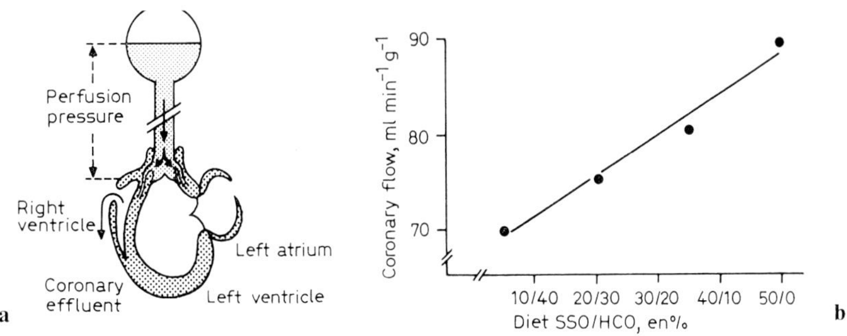

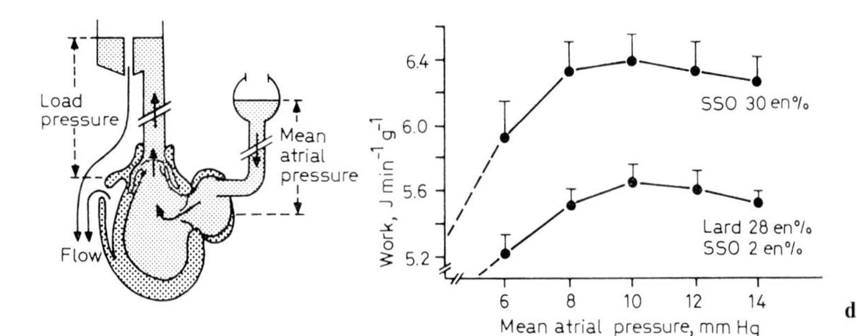

Fig. 6. Influence of dietary fats on coronary flow and myocardial contractility of the isolated rat heart. **a** In the Langendorff heart, the aortic valve is closed by the perfusion pressure and the heart muscle is perfused via the coronary arteries; the left ventricle is essentially empty. The perfusion fluid leaves the heart via the pulmonary artery and drips from the heart, after which coronary flow can be determined. Hearts are paced at a frequency of 360 min⁻¹. **b** After feeding diets (40 en% fat) with mixtures of sunflowerseed oil (SSO; 65% C 18:2) and hardened coconut oil (HCO, no C 18:2) for 3 days or longer to groups of rats (n = 8), the coronary flow is increased with increasing amounts of dietary sunflowerseed oil; flow is in ml/min/g dry heart weight. **c** In the isolated, working heart the heart is perfused via the left atrium, and the left ventricle pumps the perfusion fluid against a fixed resistance (load pressure). Left ventricular work can be calculated from flow and pressure. By changing left atrial filling pressure, left ventricular function curves can be obtained. Hearts are paced at a frequency of 360 min⁻¹. **d** Average function curves of two groups (n = 10) of rats fed diets containing 30 en% SSO (65% C 18:2) or lard (8% C 18:2) for 6 weeks. It is clearly shown that at the same mean atrial pressure the hearts of the rats fed SSO perform more work than the hearts of rats fed lard, i.e. their capacity to perform work especially at higher mean atrial pressures is higher. The vertical bars indicate the SEM.

was low (about 130/80) and the diet with the lowest P/S ratio still contained 7 en% linoleic acid; data on salt intake are not available.

It is concluded that dietary linoleic acid has a profound influence on salt-induced hypertension in rats. This is most probably partly exerted via the prostaglandin pathway. In the clinical situation, there are indications that a linoleic-acid-rich diet has a beneficial effect on 'borderline' hypertension of unknown origin. However, better designed and controlled clinical experiments are necessary to elucidate this very important aspect of dietary influence on disease.

Coronary Flow and Heart Muscle Function

Feeding rats mixtures of sunflowerseed oil and hardened coconut oil for 3 days or longer results in changes in the coronary flow of their isolated perfused hearts [20]. Figures 6a, b summarize the results of these experiments, showing that increasing the amount of dietary linoleic acid leads to an increase in coronary flow.

In the isolated, working heart of rats fed various diets, a linoleic-acid-rich diet leads to a shift to the left of the left ventricular function curve as compared to that of hearts of rats fed lard or coconut oil [21]. This shows that linoleic acid improves the contractility of the heart muscle under these circumstances. Figures 6c, d show the experimental set-up and typical results of such an experiment.

Prefeeding with a linoleic-acid-rich diet thus improves coronary flow and heart muscle function of the isolated rat heart.

Function of the Arterial Vessel Wall

The function of the arterial wall with respect to blood platelet aggregation and the clotting process, recently got more background through the finding of *Moncada et al.* [22] that the vessel wall, particularly the endothelial part of it, produces and secretes the anti-blood-platelet-aggregating and vasodilating substance prostacyclin (PGI_2). This finding and results of subsequent investigations [23–26] gave rise to the hypothesis that the balance between thromboxane A_2, a blood-platelet-aggregating and vasoconstricting substance formed by blood platelets, and prostacyclin plays an important role in the control of the interaction between blood platelets and vessel wall

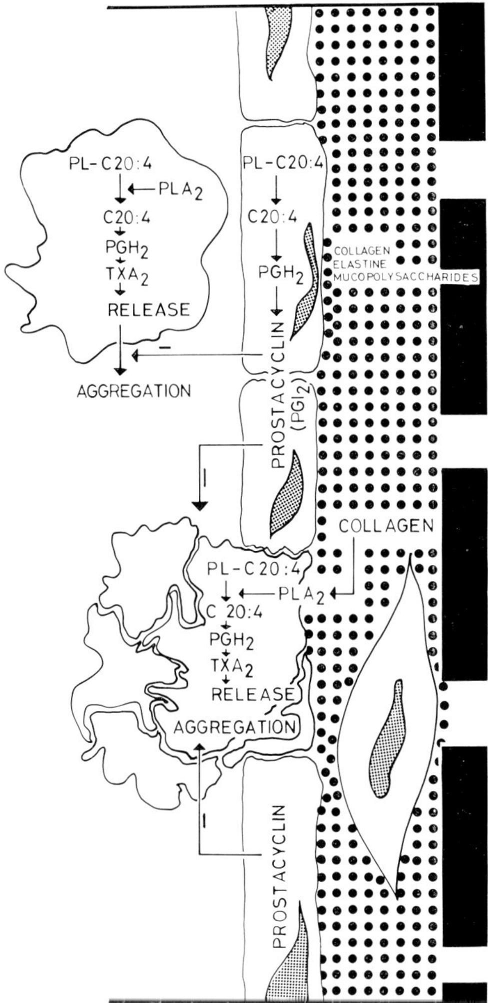

Fig. 7. Schematic representation of the interaction between blood platelets and blood vessel wall according to *Moncada et al.* [22]. When the endothelium is intact, there is a balance between prostacyclin (PGI$_2$) formation by the vessel wall and thromboxane A$_2$ (TXA$_2$) production by blood platelets. This balance can be disturbed when the endothelium is damaged.

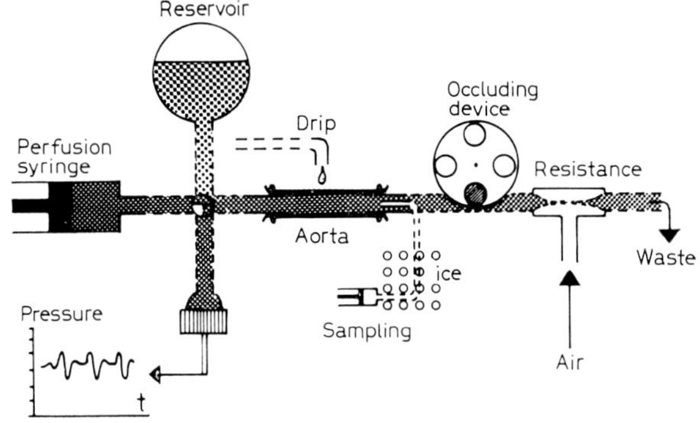

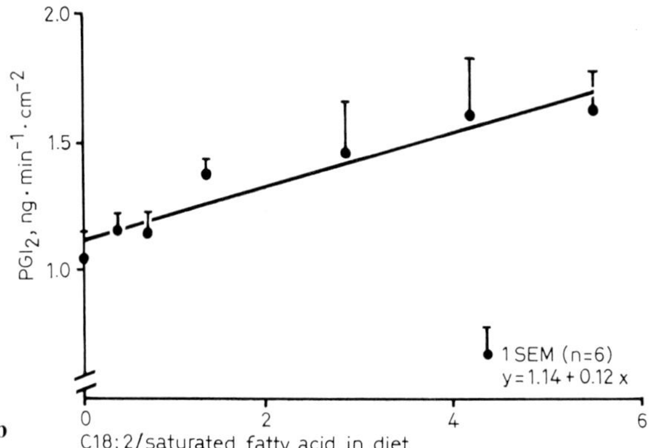

Fig. 8. a A piece of rat abdominal aorta is cannulated on both sides and (f = 300 min⁻¹) pulsation-perfused with a Krebs-Henseleit buffer solution (flow rate 0.375 ml·min⁻¹; T = 37°C; pressure variations 140/90 mm Hg). Samples are taken and immediately assayed for PGI_2 by determining the degree of inhibition of ADP-induced rat blood platelet aggregation. **b** Prostacyclin (PGI_2) production of the isolated pulsation-perfused rat aorta after feeding groups of 6 rats mixtures of hardened coconut oil and sunflowerseed oil (total 35 en%) for 4 weeks.

(fig. 7). We investigated whether diets containing various amounts of linoleic acid had an influence on the prostacyclin production of the isolated perfused rat aorta [27]. Figure 8 shows the method used and the effect of feeding mixtures of sunflowerseed oil and hardened coconut oil for 4 weeks on the prostacyclin production of the arterial wall. There is an essentially linear relationship between the dietary ratio linoleic acid/saturated fatty acids and prostacyclin production.

Discussion

The results presented show that apart from its blood cholesterol lowering effect, dietary linoleic acid has a number of well-defined cardiovascular effects which are not the consequence of abolishing a state of essential fatty acid deficiency, as defined by the well-known biochemical or biological criteria, because most of them have been found to be 'dose-dependent', i.e. increasing the amount of linoleic acid in the diet leads to increasing effects (fig. 1, 6a, b, 8). The observed phenomena are often most pronounced under circumstances in which the organ or organ system is under 'stress'. This is very clearly shown with respect to the blood pressure effect in the rat: blood pressure is kept low or is lowered by dietary linoleic acid only during salt loading and no effect of any of the diets studied is observed when this 'stress' situation is not present (fig. 4, 5). In the isolated, working, rat heart, the effect of dietary linoleic acid on heart muscle function is most pronounced at high left atrial filling pressures, i.e. under circumstances of a high load for the heart muscle (fig. 6c, d). These observations suggest important functions for linoleic acid, especially under extreme circumstances. As it is easily stored in fat depots of the body, a 'buffer amount' of linoleic acid in these depots may thus be beneficial under circumstances of 'stress'.

The multitude of physiological effects of dietary linoleic acid makes it tempting to suppose that there is a unifying underlying concept which might explain all findings. *Thomasson* [28] suggested already in 1969 that the effects of dietary linoleic acid may be explained via the formation of prostaglandins. Since then the knowledge in this field has rapidly expanded. The most important prostaglandins, thromboxanes and hydroxy fatty acids synthesized in the body from arachidonic acid, together with a simplified diagram of their metabolic pathways, are shown in figure 9. For dihomo-γ-linolenic acid, a similar diagram can be given. However, PGI cannot be formed in this pathway, as the second double bond in the side chains is essential for its formation.

Table II gives a survey of the prostaglandins, thromboxanes and hydroxy fatty acids derived from linoleic acid, and their activities which may be important with respect to the interaction between blood platelets and vessel wall and in blood pressure regulation.

These data clearly show that the aggregating agent thromboxane A_2

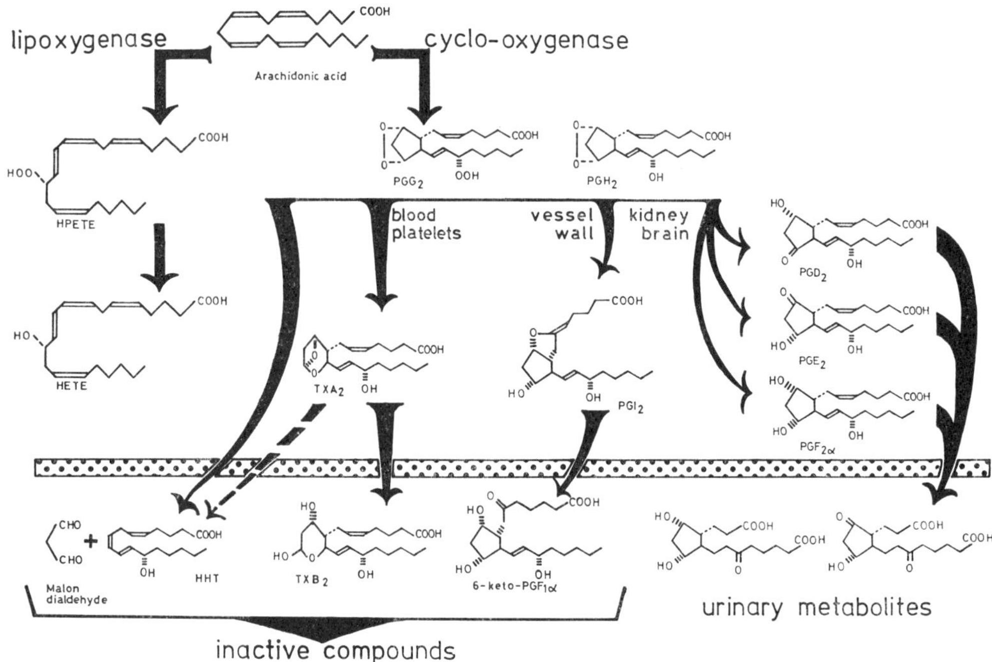

Fig. 9. Simplified diagram of arachidonic acid metabolism.

(TXA_2) as well as the anti-aggregating prostaglandins PGE_1, PGD_2 and prostacyclin (PGI_2) all can be formed from the ultimate dietary precursor linoleic acid. The cardiovascular effects of dietary linoleic acid described thus cannot simply be explained by a stimulating effect on prostaglandin synthesis. Most probably, there exist complicated regulation systems which determine the type and amount of the various prostaglandins formed for each process where prostaglandins are involved. Figure 10 schematically summarizes some of the possible interactions between the from linoleic acid derived prostaglandins with respect to the control of arterial thrombosis, atherogenesis and blood pressure.

Until now, the complexity of the systems involved makes it impossible to completely unravel the mechanism of action behind the physiological effects of dietary linoleic acid. *Van Evert et al.* [29] described a positive relationship between the amount of linoleic acid in the diet and urinary excretion of prostaglandin derivatives (fig. 11), indicating that there is a de-

Table II. Linoleic-acid-derived prostaglandins, thromboxanes and hydroxy fatty acids, their site of formation and activity in blood platelets, the cardiovascular system and kidney

Direct precursor	Prostaglandin thromboxane hydroxy fatty acid	Main site of formation	Activity	
			blood platelets	heart, blood vessels, kidney
Dihomo-γ-linolenic acid	PGG_1, PGH_1 TXA_1 PGD_1 PGE_1, $PGF_{1\alpha}$	small amounts will be formed in most organs	*not* aggregating PGE_1: strongly anti-aggregating	? PGE_1: vasodilating
Arachidonic acid	PGG_2, PGH_2	almost every cell	aggregating?	?
	TXA_2	mainly blood platelets	strongly aggregating	vasoconstricting
	PGI_2	vessel wall, lung, heart, kidney, stomach	strongly anti-aggregating	vasodilating promotes natriuresis and diuresis
	PGD_2, PGE_2, $PGF_{2\alpha}$	PGD_2: largely unknown PGE_2 and $PGF_{2\alpha}$: in kidney and brain	PGD_2: strong anti-aggregation PGE_2: slightly aggregating? $PGF_{2\alpha}$: neutral	PGE_2: vasodilating; – promotes natriuresis and diuresis $PGF_{2\alpha}$: inhibits renin secretion; veno-constricting; positive inotropic
	12-hydroxy eicosa tetraenoic acid	blood platelets	formation probably necessary for irreversible aggregation	?

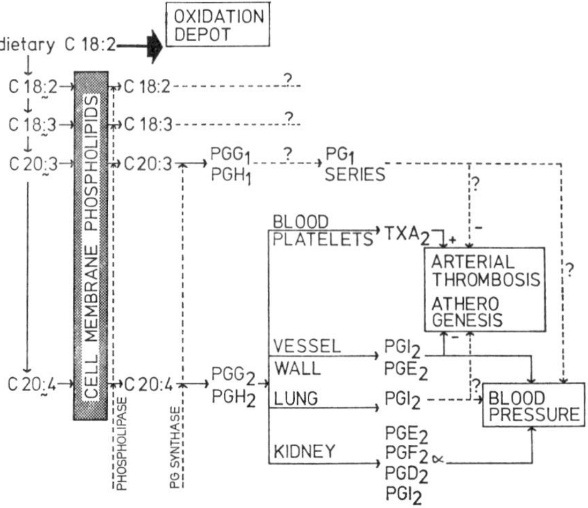

Fig. 10. Suggested interactions between linoleic acid-derived prostaglandins in the control of arterial thrombosis, atherogenesis and blood pressure.

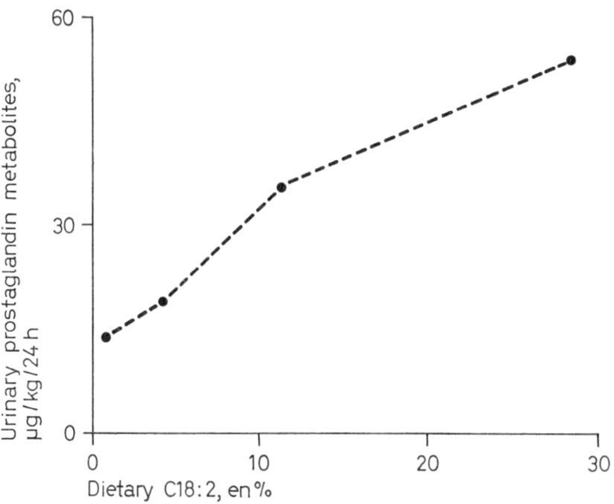

Fig. 11. Influence of dietary linoleic acid on urinary prostaglandin metabolites in rats. Diets contained 40 en % fat, consisting of mixtures of sunflowerseed oil and hardened coconut oil. Each diet group consisted of 8 rats. The measuring points represent the mean prostaglandin metabolite level during 4 months of feeding.

finite influence of dietary linoleic acid on prostaglandin metabolism. The observation that dietary linoleic acid stimulates the production of PGI_2 by the pulsation-perfused rat aorta (fig. 8) indicates that prostaglandins are involved in the effects of dietary linoleic acid on arterial thrombosis, atherogenesis and blood pressure. During salt loading, a linoleic-acid-rich diet results in the excretion of a much higher amount of urinary prostaglandin derivatives as compared to a diet poor in linoleic acid (fig. 5), indicating that at least part of the blood-pressure-lowering effect of dietary linoleic acid can be explained by an influence on prostaglandin synthesis. On the contrary, in the isolated rat heart it could not be shown that dietary linoleic acid had an appreciable influence on the release of the prostaglandin 2-series [30] and the stimulating effect of dietary linoleic acid on PGI_2 production of the aorta could not be confirmed when using a method in which pieces of rat aorta are incubated [31] or when the aorta is not pulsation-perfused [30].

Although the amount of dihomo-γ-linolenic acid in the cell membrane is small as compared to arachidonic acid, it cannot be completely excluded that prostaglandins of the 1-series play a physiological role in the body.

Apart from their function as prostaglandin precursors, the polyunsaturated fatty acids synthesized from linoleic and linolenic acid itself have an important structural function with respect to the fluidity of the cell membrane [32]. As fluidity is a very important determinant for the function of various membrane-bound enzyme systems, among them phospholipase A_2, this structural function of linoleic-acid-derived fatty acids may also be of key importance in the explanation of the observed physiological effects of dietary linoleic acid.

We conclude that, although the evidence is conflicting, the physiological effects of dietary linoleic acid can partly be explained by an influence on prostaglandin synthesis. However, this influence is complex, and most of the pathways are not known. The influence of dietary linoleic acid on membrane fluidity suggests that there is another important mechanism which may explain some of the physiological effects observed.

References

1   Burr, G.O. and Burr, M.M.: A new deficiency disease produced by rapid exclusion of fat from the diet. J. biol. Chem. *82:* 345–367 (1929).
2   Holman, R.T.: The ratio of trienoic:tetraenoic acids in tissue lipids as a measure of essential fatty acid requirement. J. Nutr. *70:* 405–410 (1960).
3   Deuel, H.J.: The effect of fat level of the diet on general nutrition. VIII. The essential

fatty acid content of margarines, shortenings, butters and cottonseed oil as determined by a new biological assay method. J. Nutr. *45:* 535–549 (1951).

4   Thomasson, H.J.: Biological standardization of essential fatty acids (a new method). Int. Z. VitamForsch. *25:* 26–82 (1953).

5   Crawford, M.A.; Hassam, A.G., and Rivers, J.P.W.: Essential fatty acid requirements in infancy. Am. J. clin. Nutr. *31:* 2181–2185 (1978).

6   Jackson, R.L.; Taunton, O.D.; Morrisett, J.D.; Gotto, A.M., Jr.: The role of dietary polyunsaturated fat in lowering blood cholesterol in man. Circulation Res. *42:* 447–453 (1978).

7   Hornstra, G. and Vendelmans, A.: Introduction of experimental arterial occlusive thrombi in rats. Atherosclerosis *17:* 369–382 (1973).

8   Hornstra, G.: Dietary fats and arterial thrombosis. Haemostasis *2:* 21–52 (1974).

9   Hornstra, G. and ten Hoor, F.: The filtragometer: a new device for measuring platelet aggregation in venous blood of man. Thromb. Diath. haemorrh. *34:* 531–544 (1975).

10 Miettinen, M.; Turpeinen, O.; Karvonen, M.J.; Elosuo, R., and Paavilainen, E.: Effect of cholesterol-lowering diet on mortality from coronary heart disease and other causes. A twelve-year clinical trial in men and women. Lancet *ii:* 835–838 (1972).

11 Hornstra, G.; Lewis, B.; Chait, A.; Turpeinen, O.; Karvonen, M.J., and Vergroesen, A.J.: Influence of dietary fat on platelet function in men. Lancet *i:* 1155–1157 (1973).

12 Triebe, G.; Block, H.U. und Förster, W.: Über das Blutdruckverhalten kochsalzbelasteter Ratten bei unterschiedlichem Linolsäuregehalt des Futters. Acta biol. med. germ. *35:* 1223–1224 (1976).

13 Ten Hoor, F. and Van de Graaf, H.M.: The influence of a linoleic-acid-rich diet and of acetyl salicylic acid on NaCl-induced hypertension, Na$^+$- and H_2O balance and urinary prostaglandin excretion in rats. Acta biol. med. germ. *37:* 875–877 (1978).

14 Dunn, M.J. and Hood, V.L.: Prostaglandins and the kidney. Am. J. Physiol. *233:* F169–F184 (1977).

15 Nugteren, D.H.: The determination of prostaglandin metabolites in human urine. J. biol. Chem. *250:* 2808–2812 (1975).

16 Iacono, J.M.; Marshall, M.W.; Dougherty; R.M.; Wheeler, M.A.; Mackin, J.F., and Canary, J.J.: Reduction in blood pressure associated with high poly-unsaturated fat diets that reduces cholesterol in man. Prev. Med. *4:* 426–443 (1975).

17 Iacono, J.M.; Marshall, M.W.; Mackin, J.F.; Dougherty, R.M.; Kliman, P.; Weinland, B.T., and Canary, J.J.: Dietary reduction of blood pressure associated with an increased urinary excretion of sodium and potassium in man. Abstract. 11th Int. Congr. Nutr., Rio de Janeirc 1978, p. 101.

18 Comberg, H.U.; Heyden, S.; Hames, C.G.; Vergroesen, A.J., and Fleischman, A.I.: Hypotensive effect of dietary prostaglandin precursor in hypertensive man. Prostaglandins *15:* 193–197 (1978).

19 The National Diet-Heart Study. Final report, monogr. No. 18 (American Heart Association, New York 1968).

20 De Deckere, E.A.M. and Ten Hoor, F.: Influences of dietary fats on the coronary flow and oxygen consumption of the isolated rat heart; in Harris, Bing and Fleckenstein, Recent advances in studies on cardiac structure and metabolism, vol. 7, pp. 475–483 (University Park Press, Baltimore 1976).

21 De Deckere, E.A.M. and Ten Hoor, F.: Effects of dietary fats on coronary flow rate and the left ventricular function of the isolated rat heart. Nutr. Metab. *23:* 88–97 (1979).

22 Moncada, S.; Gryglewski, R.J.; Bunting, S., and Vane, J.R.: An enzyme isolated from arteries transforms prostaglandin endoperoxides to an unstable substance that inhibits platelet aggregation. Nature, Lond. *263:* 663–665 (1976).

23 De Deckere, E.A.M.; Nugteren, D.H., and Ten Hoor, F.: Prostacyclin is the major prostaglandin released from the isolated perfused rabbit and rat heart. Nature, Lond. *268:* 160–163 (1977).

24 Moncada, S.; Higgs, E.A., and Vane, J.R.: Human arterial and venous tissues generate prostacyclin (prostaglandin X), a potent inhibitor of platelet aggregation. Lancet *i:* 18–20 (1977).

25 Gryglewski, R.J.; Bunting, S.; Moncada, S.; Flower, R.J., and Vane, J.R.: Arterial walls are protected against deposition of platelet thrombi by a substance (prostaglandin X) which they make from prostaglandin endoperoxides. Prostaglandins *12:* 685–713 (1976).

26 Editorial: Circulating prostacyclin. Lancet *ii:* 21–22 (1978).

27 Quadt, J.F.A. and Ten Hoor, F.: Prostacyclin synthesis of the isolated perfused aorta is stimulated by pulsations, inhibited by aspirin and nicotin and can be modulated by the type of dietary fat. Proc. 20th Dutch Fed. Meet., Groningen 1979.

28 Thomasson, H.J.: Prostaglandins and cardiovascular diseases. Nutritio Dieta *11:* 228–240 (1969).

29 Van Evert, W.C.; Soeting, W.J.; Spuy, J.H., and Nugteren, D.H.: The effect of different amounts of linoleic acid in the diet on the excretion of urinary prostaglandin metabolites in the rat. Proc. 20th Dutch Fed. Meet., Groningen 1979.

30 De Deckere, E.A.M.; Nugteren, D.H., and Ten Hoor, F.: Influence of type of dietary fat on the prostaglandin release from isolated rabbit and rat hearts and from rat aortas. Prostaglandins (in press).

31 Hornstra, G.; Haddeman, E., and Don, J.A.: Some investigations on the role of prostacyclin in thromboregulation. Thromb. Res. *12:* 367–374 (1978).

32 Houtsmuller, U.M.T.: Biochemical aspects of fatty acids with trans double bonds. FetteSeifenAnstrMittel *80:* 162–170 (1978).

Dr. F. ten Hoor, Unilever Research Vlaardingen, Olivier van Noortlaan 120, PO box 114, NL-3130 AC Vlaardingen (The Netherlands)

Discussion

Dr. Crawford: Is it reasonable to draw conclusions about prostaglandin synthesis in relation to a specific function, such as blood pressure, from assay of urinary metabolites? Is it not the case that current methodology does not differentiate between prostaglandin types of the 1 or 2 series?

Dr. ten Hoor: The urinary metabolites we determined derive from prostaglandins E and F and thus give an impression of the production of these prostaglandins in the whole body. It is not possible to differentiate between prostaglandins of the 1 and 2 series. Nevertheless, the measurements give an impression of the effect of dietary changes on prostaglandin production.

Nutr. Metab. *24* (Suppl. 1): 181–183 (1980)

Food Pattern and Mortality in Belgium[1,2]

J.V. Joossens

Division of Epidemiology, School of Public Health, University of Leuven, Leuven

Summary

In this study, the reliability of the regional nutritional data and of reported mortalities has been tested. The food data were derived from an analysis of 153 food items of the family budget enquiry of the NIS (1973–1974).

From the evidence reported and from all the observaṭions made previously by scientists in five Belgian universities, the following can be concluded. In Belgium, the most important difference between north and south in terms of risk factors of atherosclerosis is serum cholesterol and this difference can be explained by the observed differences in fat intake.

More saturated, less polyunsaturated fat and somewhat more food cholesterol are consumed in the south. This leads to a 11.9 mg/dl estimated difference in serum cholesterol according to the Keys formula. This estimated difference can be dissected into its three components. 62% of the difference comes from saturated fat, 27% from polyunsaturated and 11% comes from food cholesterol. The saturated fat difference can almost entirely be explained by the difference in butter intake. The polyunsaturated fat difference comes mostly from the higher intake of soft, salt free margarine and of corn oil. All other food ingredients are very similar except bread, cereals and salt intake which are somewhat higher in the north. More specifically, total sucrose and crude fibre are practically identical, so are the reported blood pressure, serum triglycerides, height, weight and blood groups. Smoking habits are not very different, whereas alcohol consumption is higher in the south.

In terms of coronary mortality rates in the 45- to 64-year age group, evidence has been presented that the differential mortality is not due to underclassification in the north. On the contrary, the real difference is probably greater than the reported, due to underclassification in the south.

[1] The entire article has been published in 'Colloque organisé par l'Académie Royale de Médecine de Belgique et consacré aux acides gras polyinsaturés du 14 octobre 1978'. Acta cardiol,. Suppl. 23, pp. 133–159 (1979).

[2] Supported by grants of the NFWO, Brussels.

The major mortality differences between the regions are in the coronary, cardiovascular and all-causes groups. Lung cancer is identical and so is total cancer, but stomach cancer is higher in the north. Diabetes mortality is higher in the south, but is decreasing since 1968 for both sexes together and in both regions. Butter intake is the major difference between north and south, all other factors being more or less equal. This explains why butter intake correlates so well with mortality from all causes in the belgian provinces.

The importance of genetic factors is minimised by the observation of a decreasing time trend in both sexes, for cardiovascular and total mortality, in Belgium and its regions from 1968 to 1977. This goes together with a decreasing butter consumption in both regions. A similar decrease of cardiovascular mortality is noted in the USA, Australia and Finland, but not so in 17 other European or East-European countries; at least for the now available years (1968–1974–1976).

In conclusion, it can be said that the presented evidence is entirely consistent with the diet-heart hypothesis. Further efforts should be made to lower the saturated fat intake at the population level to the FAO criteria (1977) namely 10% of calories, and to increase monounsaturated and polyunsaturated fat to the same level. This would mean on the average ± 30 g/day of each. Ideally, the final aim would be, respectively, 5, 10 and 10% and this from birth on. Food cholesterol intake should be lower than 300 mg/day for adults. From the mortality trends observed in 17 western and East-European countries, it can safely be deduced that coronary and cardiovascular mortality will remain at the same high level if food patterns are not drastically changed in those countries.

Dr. J.V.Joossens, Division of Epidemiology, School of Public Health,
University of Leuven. Kliniek St. Rafael, Kapucijnenvoer 33, B-3000 Leuven (Belgium)

Discussion

Dr. Epstein: If you compare the South of Belgium and the North of France, what do you find?

Prof. Joossens: The North of France is very similar in terms of food habits to the South of Belgium. The only difference is that the high butter intake in the North of France is combined with a high margarine intake. Mortalities in the two regions are similar and not so far away from those seen in Finland.

Dr. Crawford: Is it suitable to use lung cancer as a form of statistical control for heart disease when the incubation period for cancer and its potential for reversibility once initiated are likely to be different from those in CHD?

Prof. Joossens: It is quite probable that incubation period and potential for reversibility in lung cancer and coronary mortality are different. The point is, however, that there is no uniform pattern of behavior. In England and Wales lung cancer decreases in males whereas CHD increases. In the USA and Belgium the opposite occurs. But in Finland lung cancer decreases together with CHD.

Dr. Epstein: The decline in CHD mortality in the USA is more marked than in Belgium. If there was no change in blood pressure or in smoking in Belgium as compared to the USA, the difference could be explained in terms of changes in dietary habits only. But is there any evidence for change in blood pressure in Belgium?

Prof. Joossens: Up to now there is no direct evidence for a decrease in blood pressure in the population in Belgium. Indirect evidence is provided by the behavior of stroke, which is decreasing faster since 10 years than in neighboring countries.

Nutr. Metab. *24* (Suppl. 1): 184–186 (1980)

Structure of Fat Consumption in Hungary and Role of Interpersonal Relations in Its Modification

Eva Kurucz and J. Duba

Research Institute for Vegetable Oil and Detergent Industry, Budapest

Key Words. Fat consumption · Interpersonal relations · Animal fats · Vegetable fats · Cholesterol levels

Abstract. In Hungary, fat consumption is inadequate both from a quantitative and from a qualitative point of view. Statistical data also reveal that *per capita* consumption of fat has become stabilized and that within the vegetable fats/oils, margarine consumption increases year by year. To accelerate this process, a study was carried out on the effect of personal persuasion on the consumption of vegetable oils and different kinds of margarine. The results of this investigation are summarized in our paper.

Fat consumption in Hungary is inadequate both from a quantitative and from a qualitative point of view: the *per capita* consumption of fat is too high and the majority of the population consumes fat of mostly animal origin. On the other hand, however, statistics on food consumption reveal that the fat consumption per person has stabilized around 28.5 kg per year; quality vegetable fat/oil and margarine consumptions are increasing (5.2% in 1960, 18.1% in 1977; table I). Fat consumption thus follows a biologically satisfactory course. To accelerate this process, we carried out studies in collaboration with the Cardiological Institute, Budapest, on the effect of personal persuasion on dietary habits, namely to what extent a doctor can convince his patient to consume vegetable oils and different kinds of margarine instead of animal fats.

We aimed at determining the dietary habits of the population of Budapest and the percentage of fat and oil in their meals. We wanted to find out what percentage of those who consumed fat of animal origin would be willing to consume fats of vegetable origin as a consequence of personal

Table I. Evolution of the fat consumption per person in Hungary from 1960 to 1977

	1960	1970	1975	1976	1977
Total fat intake, kg	23.1	27.1	28.5	28.3	28.7
Total fat intake, %	100.0	117.8	123.9	123.0	124.8
Vegetable oil	0.8	1.9	2.9	3.3	5.2
Margarine[1]	0.4	0.7	1.4	1.6	
Butter[1]	1.1	1.7	1.4	1.4	1.4
Lard	19.8	22.0	22.1	21.2	
Poultry fat	1.0	0.8	0.7	0.8	22.1
Animal fats, %	94.8	90.4	84.9	82.7	81.9
Vegetable fats, %	5.2	9.6	15.1	17.3	18.1

[1] Butter and margarine are expressed in values equivalent to fat values.

talks and information leaflets handed over to them and also as a result of dietetic advice. We also studied to what extent changing over to vegetable fats influenced the serum cholesterol values.

Our studies were performed on men between 35 and 65 years of age whose serum cholesterol values fell into the upper or lower third of the cholesterol curve for the healthy population.

Dietary habits were determined with questionnaires, and cholesterol levels were determined according to the Watson method. The questionnaires gathered information on the number and of meals and the place where they are taken (work, restaurant, at home), the quantity of fat used for preparing the meals as well as the amount of butter, margarine, egg and bacon consumption. Those who mainly consumed fats of animal origin were asked whether they would be willing to change over to consuming fats of vegetable origin.

In the low-cholesterol group, we studied 470 persons; their average cholesterol value was 204 ± 39 mg%. The high-cholesterol group consisted of 962 persons; their average cholesterol value was 250 ± 43 mg%.

According to our investigations, 35% of the participants of the test used only oil for cooking, 46% used only lard and 18% prepared their meals with oil and fat alike. There was no significant difference between the groups with low or high cholesterol values. From among those who used lard for cooking, 37% declared their willingness to change over to using oil and margarine, 29% said that they would do so if their wives agreed; 34% gave a definitely negative answer.

Control examinations will be carried out in the near future and we hope they will prove that personal discussions with the doctor and information leaflets handed over simultaneously are the most effective ways of amending fat consumption habits and of bringing about healthy dietary habits.

Dr. Eva Kurucz, Research Institute for Vegetable Oil and Detergent Industry, Maglódi ut 6, 1106 Budapest (Hungary)

Nutr. Metab. *24* (Suppl. 1): 187–199 (1980)

Contaminations et réactions secondaires dans la fabrication des corps gras

J. P. Wolff

Institut des Corps Gras, Paris

Key Words. Contaminants · Secondary reactions · Processing

Abstract. The contamination of dietary fats and oils can have various origins: contamination of the raw materials that could not be eliminated by the industrial process; refining completely eliminates this type of contamination; contamination with one of the processing aids can occur if the process is not properly conducted; contamination with packaging materials will increase with the storage duration of the finished product (in the case of vinyl chloride monomer and styrene contamination); only the strict control and adequate choice of the materials and packaging conditions will allow complete elimination of this contamination type. The presence of small amounts of another fat (in vegetable margarines for instance) should also be mentioned although it is not a real contamination. The secondary reactions that may occur during the processing of dietary fats are primarily due to unwanted chemical reactions leading to the formation of fatty acid or native glyceride isomers (e.g. formation of geometric isomers during deodorization, formation of position isomers during hydrogenation). In the present stage of knowledge, it is fairly easy to avoid secondary reactions during deodorization, those that take place during hydrogenation are not yet under control, however.

Quel que soit le soin apporté aux fabrications, les industries alimentaires ne peuvent éviter d'être confrontées à des risques de contamination ou de réactions secondaires dont l'élimination totale pose parfois des problèmes technologiques ou scientifiques difficiles à résoudre. Dans le domaine des corps gras, les origines des contaminations éventuelles sont assez bien définies pour qu'il soit possible de les éliminer complètement d'une façon presque générale. Par contre, la chimie des corps gras est trop complexe pour que l'on puisse dans tous les cas éviter, lors des traitements technologiques, des réactions parasites qui, même si elles sont sans inconvénients majeurs, ne sont pas toujours désirées par le producteur.

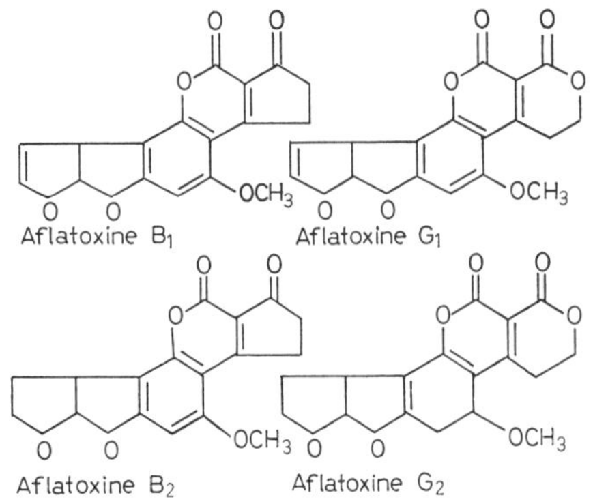

Fig. 1. Structure des aflatoxines B_1, B_2, G_1 et G_2.

Contaminations

La contamination des corps gras alimentaires par des produits étrangers peut avoir trois origines principales.

La contamination des matières premières

Le cas le plus connu et le plus étudié est celui des aflatoxines secrétées par l'*Aspergillus flavus*.

La toxicité aiguë des aflatoxines (fig. 1) est considérable. Les doses léthales 50 sont, pour le caneton, les suivantes: B_1: 0,4 mg/kg (ce qui correspond à 28 mg pour un homme adulte); B_2: 1,7 mg/kg; G_1: 1,2 mg/kg; G_2: 3,4 mg/kg. D'une façon générale, la sensibilité des animaux de laboratoire, des animaux d'élevage et des êtres humains dépend de leur âge et les jeunes sont moins résistants que les adultes. L'intoxication chronique par les aflatoxines conduit chez toutes les espèces à l'apparition de cancers du foie.

Les huiles brutes provenant de graines contaminées sont elles-mêmes contaminées. En raison de leur structure, les aflatoxines sont peu solubles dans les huiles et généralement la contamination des huiles est cinq fois inférieure à celle des graines dont elles proviennent. Elle ne dépasse que de façon tout à fait exceptionnelle 100 ppb (c'est-à-dire 100 mg/t). Quelle que soit cette concentration, les opérations classiques du raffinage des huiles alimentaires conduisent à leur élimination totale.

La soude utilisée pour la neutralisation réagit sur la fonction lactone des aflatoxines et en détruit 90–98%. Le reste est absorbé par les terres dé-

Tableau I. Influence du raffinage sur la teneur en insecticides organochlorés des huiles

Nature de l'huile	Insecticides, ppm			
	HCH	heptachlore	Aldrine	DDT+DDE+DDD
Colza				
Brute	10,2	10,8	11,1	34,6
Neutralisée	10,2	10,8	11,1	34,6
Décolorée	10,2	10,8	11,1	34,6
Désodorisée	0,005	–	–	0,024
Tournesol				
Brute	9,3	10,4	10,4	31,7
Neutralisée	9,3	10,4	10,4	31,7
Décolorée	9,3	10,4	10,4	31,7
Désodorisée	0,009	–	–	0,016

Les teneurs moyennes en insecticides organochlorés des huiles brutes ne dépassent normalement pas 0,2 ppm pour l'HCH et 0,01 ppm pour chacun des autres insecticides.

colorantes utilisées pour «blanchir» les huiles. Leur élimination dans l'huile raffinée est donc totale.

Un autre exemple, comparable, est celui de la présence de résidus d'insecticides organochlorés ou organophosphorés dans les huiles et graisses provenant de produits végétaux traités par ces insecticides trop peu de temps avant leur entrée en huilerie, ou de tissus adipeux d'animaux nourris avec des aliments (entre autres des tourteaux) contaminés.

Les dérivés organochlorés (HCC, Dieldrine, Aldrine, DDT), dont l'utilisation est de plus en plus strictement réglementée, sont dans l'ensemble des corps chimiquement très stables et assez solubles dans les huiles et les graisses. Si les matières premières mises en œuvre sont contaminées, les produits gras obtenus le seront également. En raison de la grande stabilité chimique et de la polarité relativement faible des insecticides organochlorés, la plupart des opérations de raffinage sont sans action notable sur leur concentration dans le corps gras. Mais, comme tous ces corps sont relativement volatils, la désodorisation par la vapeur d'eau à des températures supérieures à 180–200 °C permet l'élimination totale de ce type de polluant quelle que soit sa concentration initiale dans l'huile brute, comme le montrent les résultats du tableau I relatifs à des huiles volontairement enrichies en insecticides organochlorés.

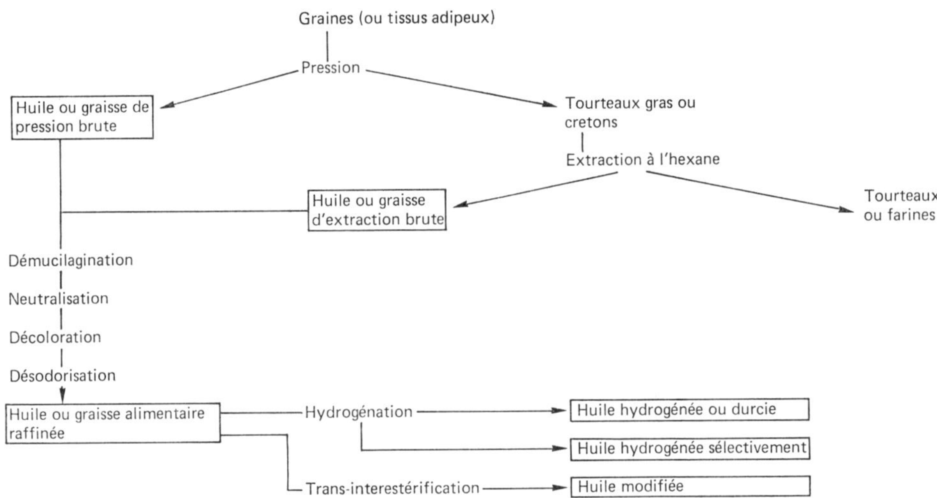

Fig. 2. Schéma du procédé d'obtention des corps gras alimentaires.

La très faible résistance des insecticides organophosphorés (malation, parathion) à l'hydrolyse et en particulier à l'hydrolyse en milieu alcalin permet également au raffinage de les éliminer totalement.

La contamination par les processus technologiques

Peut-être est-il nécessaire de rappeler très succintement le schéma de l'obtention des huiles et graisses alimentaires (fig. 2).

La pression est un procédé qui, n'utilisant pas d'auxiliaire technologique, ne peut introduire de contaminants à l'exception, le cas échéant, de traces des métaux constituant les broyeurs et les presses.

Les sous-produits de la pression (tourteaux gras dans le cas des graines, cretons dans le cas des tissus adipeux) contiennent encore une quantité importante de corps gras (souvent plus de 12%) qu'il est économiquement nécessaire de récupérer. Ceci se fait par exemple par extraction à l'aide d'hexane (ou d'essence) à température proche de la température d'ébullition dans un appareillage entièrement clos en utilisant le plus souvent le principe du contre-courant.

En fin d'extraction, l'huile est débarrassée de l'hexane par distillation aussi complète que possible, non seulement pour des raisons économiques, mais surtout pour éviter d'envoyer au raffinage une huile contenant encore des quantités notables d'hexane (par exemple 0,5%) et qui pourrait donner, à l'occasion du chauffage, des mélanges explosifs hexane-air.

L'huile d'extraction brute contient en principe moins de 150 ppm d'hexane et en général cette valeur limite n'est pas atteinte en pratique. Il s'agit cependant, même au niveau d'une centaine de ppm, d'une contamination non négligeable et il est indispensable de savoir si le raffinage de l'huile d'extraction conduit à l'élimination du contaminant.

A chaque stade du raffinage, on observe une diminution faible de la contamination et c'est la désodorisation qui entraîne une décontamination totale. D'après notre expérience, il n'y a pas de corps gras alimentaires qui contiennent plus de 2 ppm (mg/kg) d'hexane et la quasi-totalité de la production française contient moins de 0,1 ppm d'hexane, c'est-à-dire moins de 100 mg/t. Des quantités aussi faibles étaient pratiquement indécelables il y a 10 ans et elles sont comparables aux quantités de pentane (hydrocarbure analogue à l'hexane, mais un peu plus volatil) qui se forment naturellement au cours des réactions chimiques très complexes qui constituent le rancissement des huiles.

La neutralisation des acides gras libres présents dans les huiles brutes se fait généralement par action de la soude qui transforme les acides gras libres en savons. Ceux-ci sont ensuite facilement éliminés par centrifugation et lavage.

Les corps gras bien raffinés ne contiennent pas de quantités décelables de savon par les techniques classiques, c'est-à-dire qu'elles en renferment moins de 10 ppm, et elles ne peuvent en aucun cas contenir des quantités aussi faibles soient-elles de soude libre, car celle-ci réagit sur l'huile neutre pour donner du savon.

Bien que le savon à faible concentration soit dépourvu pour l'homme de toxicité, son élimination aussi complète que possible est importante technologiquement car sa présence diminue la résistance de l'huile au rancissement.

L'élimination des acides gras libres peut se faire par une autre technique, la désodorisation désacidifiante qui entraîne les acides gras libres par la vapeur d'eau surchauffée lors d'une désodorisation conduite à plus haute température que la désodorisation simple. Dans ce cas, les problèmes de contamination par le savon ne se posent évidemment plus.

La décoloration par les terres activées qui sont des terres naturelles, rendues plus adsorbantes par lavage avec des acides minéraux (phosphorique, sulfurique, chlorhydrique) adsorbe les pigments colorés présents dans les huiles neutralisées (carotènes, chlorophylle, produits d'oxydation), la presque

Tableau II. Influence de la décoloration sur la teneur en fer et en cuivre des huiles de colza

	No. de l'échantillon							
	16	19	20	22	23	25	27	32
Teneur en fer avant décoloration	90	100	50	175	115	120	190	330
Teneur en fer après décoloration	60	10	15	20	35	40	25	130
Teneur en cuivre avant décoloration	20	15	15	18	55	26	40	25
Teneur en cuivre après décoloration	3	3	5	11	3	5	5	7
Teneur en sodium avant décoloration	300	300	300	700	700	500	400	<90
Teneur en sodium après décoloration	<90	<90	<90	<90	<90	<90	<90	<90

Les teneurs sont exprimées en ppm de métal dans l'huile.

totalité des savons de soude qui peuvent subsister après les lavages et la quasi-totalité des savons de fer ou de cuivre que les huiles peuvent renfermer (tab. II) et qui proviennent essentiellement de l'appareillage utilisé pour obtenir les huiles brutes.

La désodorisation par la vapeur d'eau sous vide à haute température ne peut introduire aucun contaminant si l'appareillage est propre et réalisé avec les matérieux adéquats.

Nous avons vu sur la figure 2 que le raffinage des corps gras alimentaires pouvait être suivi, surtout dans le cas de leur emploi en margarinerie, de traitements supplémentaires comme l'hydrogénation et l'inter- ou trans-estérification qui ont pour but de modifier les propriétés rhéologiques des corps gras et les rendre plus aptes à la fabrication des margarines.

Ces traitements utilisent des auxiliaires technologiques dont les seuls qui peuvent se retrouver dans les produits finis sont les catalyseurs: Nickel pour l'hydrogénation, sodium ou méthylate de sodium pour l'inter- ou transestérification. Le sodium ne peut rester dans une huile sous forme métallique et il est transformé en savon de sodium qui, nous l'avons vu, est complètement éliminé. La majeure partie du nickel est récupérée par filtration en vue de son réemploi ultérieur, la fraction non retenue sur les filtres étant ensuite éliminée par les traitements analogues à ceux du raffinage auxquels on soumet les huiles hydrogénées. Cette technologie est suffisamment avancée pour que la teneur en nickel des huiles hydrogénées soit à peine supérieure au seuil de détection de la méthode de dosage la plus sensible, c'est-à-dire 3 à 5 ppb. A un tel niveau, il est difficile de parler de contamination.

La contamination par les matériaux d'emballage

C'est actuellement celle qui paraît plus préoccupante, car elle est essentiellement liée à l'emploi d'emballages en matière plastique relativement récents, c'est donc une contamination moins connue que les précédentes. D'autre part, le processus technologique de fabrication de l'huile ou de la margarine est sans action sur elle puisqu'elle se produit après la préparation du produit alimentaire et qu'il n'est pas impossible de penser qu'inexistante à la sortie de l'usine, elle peut se développer dans le temps.

Il convient tout d'abord de remarquer que la contamination par l'emballage n'est pas aussi récente qu'on l'imagine en général, et que des matériaux «classiques» comme le verre ou le fer blanc pouvaient contaminer les corps gras en cours de stockage, favorisant la formation progressive de savons.

Il est cependant indéniable que l'emploi d'emballages à base de polychlorure de vinyle (PVC) et de polystyrène a posé des problèmes nouveaux essentiellement liés à la migration du chlorure de vinyle monomère et à celle du styrène monomère du matériau d'emballage dans l'aliment emballé.

Les premiers indices d'action cancérigène probable du chlorure de vinyle monomère furent mis en évidence en janvier 1974. Le problème de son dosage se posa alors avec acuité. L'amélioration des techniques de dosage permit aux producteurs d'améliorer la qualité du PVC en diminuant sa teneur en monomère.

Dès 1977 les échantillons mis en vente sur le marché français (échantillons A–H; tab. III) ne contenaient plus de quantités décelables de chlorure de vinyle monomère.

La comparaison de ces résultats avec ceux des échantillons emballés dans des chlorures de vinyle de 2 ans plus anciens montre les progrès réalisés en quelques années (Échantillons 1–5; tab. III).

Actuellement, la teneur en monomères des PVC utilisés dans le conditionnement des denrées alimentaires est si faible que le dosage direct de la migration par la technique la plus employée (chromatographie en phase gazeuse de «l'espace de tête») n'est plus possible dans les huiles sans risque d'erreurs dues à des interférences avec les constituants volatils des huiles qui

Tableau III. Teneur en chlorure de vinyle monomère (CVM) dans les huiles conditionnées dans le PVC

Echantillons	Teneur en CVM, ppb	
	(a)	(b)
1	105	118
2	40	37
3	40	23
4	100	104
5	110	144
A	$\leq$10	$\leq$5
B	$\leq$10	$\leq$5
C	$\leq$10	$\leq$5
D	$\leq$10	$\leq$5
E	$\leq$10	$\leq$5
F	$\leq$10	$\leq$5
G	$\leq$10	$\leq$5
H	$\leq$10	$\leq$5

(a) = ITERG; (b) = Service Fédéral de l'hygiène publique à Berne.

apparaissent dès les premiers stades de l'oxydation et du rancissement des corps gras.

On peut utiliser, pour le dosage direct de la migration, le couplage chromatographie en phase gazeuse/spectrométrie de masse. Mais il s'agit d'une technique encore très onéreuse et on préfère se contenter de doser le monomère dans l'emballage (sa concentration y est en moyenne inférieure à 0,2 ppm) et calculer la migration maximale possible compte tenu des poids respectifs contenu-contenant et du fait qu'une partie (environ 60%) du monomère migre à l'extérieur de la bouteille. Par ce calcul, il est possible d'affirmer qu'actuellement les corps gras emballés avec du PVC contiennent moins de 3 ppb de chlorure de vinyle monomère.

Le cas du styrène est un peu plus complexe, bien que la toxicité du styrène monomère soit très faible et que le refus, pour des raisons organoleptiques, par les consommateurs se produise pour des concentrations bien inférieures aux concentrations nuisibles à la santé. En effet, la teneur en styrène monomère du matériau d'emballage ne dépend pas que de la teneur en monomère du polystyrène, mais aussi de la technique de fabrication de l'emballage, puisque le polystyrène se dépolymérise aux températures élevées et peut former du monomère lors de la fabrication du pot. Enfin, le polystyrène est rarement employé pur mais en général dans des copolymères (ABS) ou, plus souvent, dans des «sandwichs» polystyrène-polyéthylène, polystyrène-PVC à deux ou trois couches où le polystyrène peut être au contact de la denrée alimentaire ou, au contraire, à l'extérieur.

D'après les dosages que nous avons réalisés au cours de ces derniers mois à l'ITERG sur des margarines ou des graisses emballées dans ce type de «sandwich» on peut affirmer que si le polystyrène est au contact de l'aliment, il y a migration du monomère. L'importance de la migration est extrêmement variable suivant le type d'emballage et ses conditions de fabrication. Il nous paraît cependant probable que les deux facteurs essentiels sont: la teneur en monomère du matériau de départ et le poids d'emballage par rapport à la quantité de produit emballé.

Par contre, le chlorure de polyvinyle, lorsqu'il est au contact de la denrée alimentaire constitue une barrière pratiquement infranchissable pour le styrène monomère dans les conditions normales de stockage. Dès maintenant il est possible de dire que si ce type de contamination existe et peu parfois être de l'ordre de la ppm, il est possible d'y remédier par une sélection stricte des matériaux d'emballage utilisés.

Contamination d'un corps gras par d'autres corps gras

Ce genre de contamination ne concerne que les corps gras purs, ou les mélanges pour lesquels certaines propriétés sont mises en évidence dans l'étiquetage: C'est en particulier le cas des «margarines végétales».

Les risques de contamination par des corps gras animaux, dans ce cas, se situent en deux points: contamination des matières premières et contamination des circuits de fabrication. Pour éviter la seconde, il existe une solution simple, mais onéreuse, c'est d'équiper l'usine d'un appareillage uniquement destiné à ce type de fabrication. Mais en supposant même que cela soit réalisé, on n'élimine pas pour autant la première source de contamination, fréquente dans le cas des corps gras concrets et plus particulièrement de l'huile de palme où le pourcentage d'échantillons certainement pollués par de faibles doses d'huiles marines peut être, pour certaines provenances, de l'ordre de 30%.

On peut donc dire qu'il existe là une pollution inévitable compte tenu de l'importance des unités de fabrication et de stockage et des difficultés de nettoyage des bacs de stockage et des circuits contenant des corps gras.

Le problème est donc de tenir compte de ces difficultés et de ces pollutions inévitables lors du contrôle des margarines végétales sans toutefois tolérer la présence de doses telles de corps gras animaux qu'elles pourraient correspondre à des adultérations volontaires. La détermination de la teneur en corps gras animaux d'un mélange de corps gras végétaux est un problème analytique complexe, qui n'a pas encore trouvé de solution parfaitement satisfaisante.

Cette question, très étudiée en France depuis 2 ans, soulève donc un double problème juridique et technique auquel il est difficile de trouver une solution.

Réactions secondaires

Les plus importants dans l'industrie des corps gras alimentaires sont dus à des réactions chimiques, non désirées, si ce n'est indésirables, qui donnent naissance à la formation d'isomères des acides gras ou des glycérides naturels qui n'existent pas dans le produit de départ:

Historiquement parlant, les deux plus anciennement connus sont: la formation de systèmes triéniques conjugués lors de la décoloration des corps gras par les terres activées et la formation d'isomères trans lors de l'hydrogénation partielle des huiles insaturées.

Il est bien connu qu'en utilisant des charbons actifs au lieu de terres décolorantes, on supprime la formation de systèmes polyéniques conjugués dans les huiles décolorées, mais les difficultés technologiques liées à l'emploi des charbons actifs sont telles que jusqu'à présent on continue à décolorer les huiles avec des terres. Tout au plus peut-on limiter la formation de triènes conjugués en diminuant la quantité de terres utilisées et en raffinant des huiles provenant de graines aussi peu altérées que possible.

De même, en modifiant les conditions de l'hydrogénation, ou le catalyseur utilisé, on peut limiter la formation d'isomères trans. Malheureusement, on sait aussi que si l'on diminue la formation d'isomère trans on diminue en même temps la sélectivité du catalyseur, ce qui, de façon très schématique revient à dire que l'on favorise la formation d'acides gras saturés. Malgré tous les travaux sur ce sujet, il n'est donc actuellement pas possible de dire que la technologie de l'hydrogénation évite totalement les réactions secondaires. Si l'on étudie, comme la chromatographie en phase gazeuse avec des colonnes capillaires de verre le permet, la formation des isomères de position d'un même acide insaturé, on constate que les réactions secondaires dans l'hydrogénation sont encore plus importants qu'on ne pouvait le penser il y a quelques années.

Deux autres exemples montrent, contrairement aux deux précédents, que l'approche systématique du sujet permet, dans certains cas lorsque l'on a mis en évidence les réactions secondaires liés à un processus technologique donné, de modifier celui-ci de façon à les supprimer.

C'est en 1974 qu'on a signalé, pour la première fois, la présence dans les huiles alimentaires raffinées contenant de l'acide linolénique (colza et soja) d'isomères géométriques de l'acide linolénique n'existant pas dans les des isomères géométriques de l'acide linolénique n'existant pas dans les huiles brutes de départ. En particulier les acides cis$_9$-cis$_{12}$-trans$_{15}$ octadéca-triénoïque et trans$_9$-cis$_{12}$-cis$_{15}$ octadécatriénoïque.

On pense que ces réactions secondaires sont liés à la température de désodorisation. Nous avons effectivement vérifié que l'huile de soja raffinée, mais non désodorisée ne contenait pas d'isomères trans de l'acide linolénique, et que ceux-ci se formaient au cours de la désodorisation. La quantité d'iso-mères trans prenant naissance au cours de la désodorisation est approxima-tivement proportionnelle à la température de désodorisation, et à tempéra-ture constante, proportionnelle à la durée de désodorisation. A partir d'une huile contenant 6,9% d'acide linolénique il se forme, après 6 h de désodori-sation à 200 °C, 0,8% d'isomères trans. Après 1 h de désodorisation à 240 °C il s'en est déjà formé 1,4%. La diminution ou la suppression de cette réaction secondaire (dans la limite où il serait démontré qu'elle présente des incon-vénients) passe donc par la diminution de la température de désodorisation des huiles.

L'inter- et la transestérification des corps gras sont des techniques bien connues et utilisées pour modifier les propriétés rhéologiques de ces derniers, en particulier du saindoux utilisé en biscotterie, biscuiterie, ou des corps gras ou fractions de corps gras utilisés dans la préparation de substituts de beurre de cacao. La formation de glycérides autres que ceux de l'huile de départ n'est pas une réaction secondaire puisqu'il s'agit au contraire du but pour-suivi.

On sait, depuis les travaux effectués sur l'huile d'olive en 1971 que la désacidification par la vapeur, préférée parfois à la neutralisation par la soude parce qu'elle n'emploie pas de produit chimique, s'accompagne à partir de 240 °C de phénomènes d'interestérification non négligeables, con-duisant à la formation de glycéride possédant de l'acide palmitique ou stéa-rique en position β sur le glycérol, glycéride qui n'existe pas dans les huiles d'olive n'ayant pas subi ce traitement. Le phénomène est d'autant plus important que le traitement thermique a lieu en présence de fer. Il ne faut donc pas en conclure que la désacidification «physique» s'accompagne nécessairement de réactions secondaires, mais que pour les éviter il faut limiter la température de désacidification et surtout utiliser un appareillage en bon état, n'introduisant pas de fer dans l'huile traitée. Cette conclusion d'un travail sur l'huile d'olive a pu être étendue à d'autres huiles.

Nous espérons, par ces quelques exemples, avoir montré que les moyens de la chimie analytique actuels permettent de mieux dépister les contaminations ou les modifications involontaires que peuvent subir les corps gras alimentaires. Ainsi les producteurs de corps gras sont mieux à même de fournir aux consommateurs des produits présentant le maximum de garantie du point de vue de la santé publique dans la limite où les biochimistes, les nutritionistes, les médecins et les toxicologues formulent leurs exigences avec précision.

Résumé

La contamination des huiles et des graisses alimentaires peut avoir des origines différentes: une contamination de la matière première que le processus de fabrication n'élimine pas. Le raffinage supprime complètement ce type de contamination; une contamination par le processus technologique lui-même, qui, mal conduit, laisse dans le produit fini des traces d'un des auxiliaires technologiques mis en œuvre; une contamination par le matériau d'emballage, qui risque d'augmenter avec la durée de stockage du produit fini (cas de la contamination par le chlorure de vinyle monomère et le styrène); seul le contrôle rigoureux du choix des matériaux et des conditions d'emballage permet l'élimination complète de ce type de contamination. Bien qu'il ne s'agisse pas, à proprement parler d'une contamination, il n'est pas possible de passer sous silence les problèmes soulevés par la présence dans un corps gras (par exemple les margarines végétales) de petites quantités d'un autre corps gras. Les réactions secondaires qui peuvent se produire dans la préparation des corps gras alimentaires sont essentiellement dues à des réactions chimiques non désirées conduisant à la formation d'isomères des acides gras ou des glycérides naturels (par exemple, formation d'isomères géométriques lors de la désodorisation, formation d'isomères géométriques et d'isomères de position lors de l'hydrogénation). Si dans l'état actuel de nos connaissances, il est relativement facile d'éviter les réactions secondaires lors de la désodorisation, celles qui se produisent lors de l'hydrogénation ne sont pas encore maîtrisées.

Dr Jean-Pierre Wolff, Directeur Général de l'Institut des Corps Gras,
5, boulevard de Latour-Maubourg, F-75007 Paris (France)

Discussion

Mr. Kestens: Comment peut-on éliminer le nickel des huiles hydrogénées?

Prof. Wolff: L'élimination se fait d'abord par filtration – afin de récupérer le nickel. Ensuite il y a semi-raffinage (lavage de l'huile avec des solutions alcalines, filtration et désodorisation).

Ce semi-raffinage élimine les dernières traces de nickel sous forme de savon.

Nutr. Metab. *24* (Suppl. 1): 200–210 (1979)

Heated Oils – Chemistry and Nutritional Aspects

G. Billek

Unilever Forschungsgesellschaft mbH, Hamburg

Key Words. Heated oils · Deep-fat frying · Deterioration · Health hazards · Long-term feeding experiments

Abstract. Used frying oil can be separated by means of column chromatography on silica gel into an unpolar fraction, which contains predominantly the unaltered triglycerides, and a polar fraction consisting of all oxidation and decomposition products formed during the heating process. The size of the polar fraction indicates the degree of fat deterioration. A similar procedure was applied to obtain fractions from a heated oil to be used in long-term feeding experiments. Several tons of a sunflower oil which had been used in the industrial production of fish fingers were separated into a polar fraction (1) and an unpolar fraction (2). The sunflower oil had not been overheated and was taken at the moment when the production would have been stopped, according to factory practice, and the oil discarded. Fractions 1 (group U) and fraction 2 (group P) as well as the original unheated sunflower oil (group F) and the heated sunflower oil (group H) were fed to rats over 18 months at a level of 20% in the diet. Fraction 1 caused a highly significant reduction in weight gain of the animals as compared with unheated sunflower oil but had only an insignificant detrimental effect upon the many biochemical, histological and clinical parameters. The order of the weight gain caused by the four samples was: U< H< P< F. The changes of other parameters as well as the implications of these long-term feeding studies will be discussed in detail.

Introduction

Heating changes the composition of oils and new compounds are formed. The amount of decomposition products increases gradually, the longer the oils are heated. Food legislation in many countries clearly distinguishes between deterioration and harmfulness. After a certain heating time, the used frying oil shows defects in odour and taste, and decomposition

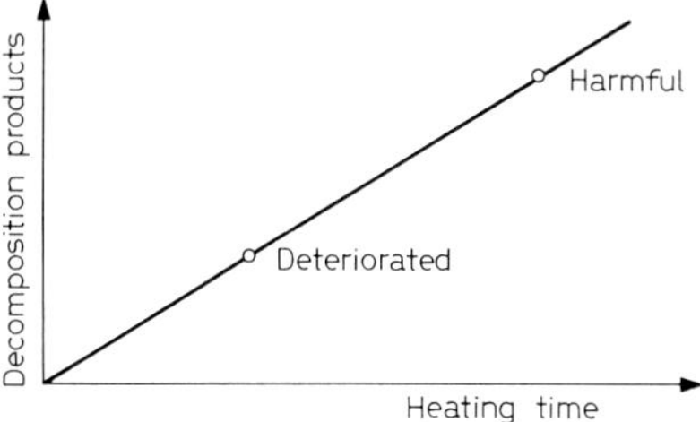

Fig. 1. Schematic diagram on the formation of decomposition products in heated oils.

products accumulate to such an extent that this oil has to be regarded as being deteriorated. Only after a further increase of the heating time might a point sometimes be reached, when the oil starts to be harmful or detrimental to health (fig. 1).

Assessment of Deterioration

Various methods have already been proposed for the assessment of the quality of used frying oils. However, within this expose, only one simple, quick and reliable method will be mentioned. This is the application of the well-known column chromatography. The oil to be analysed is brought into the column without any pre-treatment except for filtration, if necessary (fig. 2). By selecting appropriate solvents, it is possible to separate the oil in two fractions. With solvent A, we obtain the unchanged triglycerides which are relatively unpolar. With solvent B, the polar components are collected. Both fractions are evaporated and the residues are dried to constant weight. The success of this separation can easily be checked by thin-layer chromatography.

Fraction 1 (fig. 3) contains part of the unsaponifiable matter and the unaltered triglycerides. The more polar fraction, fraction 2, contains all oxidation and decomposition products formed during the heating process. The size of the polar fraction indicates the degree of fat deterioration.

Figure 4 shows how a certain fat deteriorates in a commercial fryer. The frying good was pommes frites. In this special case, after a frying time of 56 h the amount of the polar fraction reached 30%. According to our experience, in accordance with other analytical data and especially on the basis of organoleptic assessment, 30% oxidation and decomposition products indicate deterioration.

A frying oil is deteriorated if the amount of decomposition products is 30% or higher (fig. 1). Now the question arises: is it possible to correlate *harmfulness* with analytical data? To achieve this goal, a new approach was necessary. The basic idea was: take a used frying oil and separate it into two fractions, one containing oxidised material and the other the unchanged part of the used oil. Of course, the oil should not have been maltreated and the separation procedure should be as gentle as possible to exclude subsequent destruction.

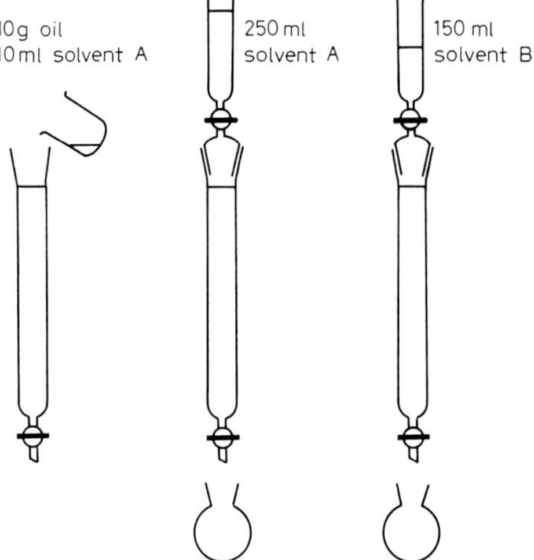

Fig. 2. Isolation of unchanged triglycerides and of polar components from heated oils. Solvent A = 80% hexane, 20% diisopropylether; solvent B = isopropanol.

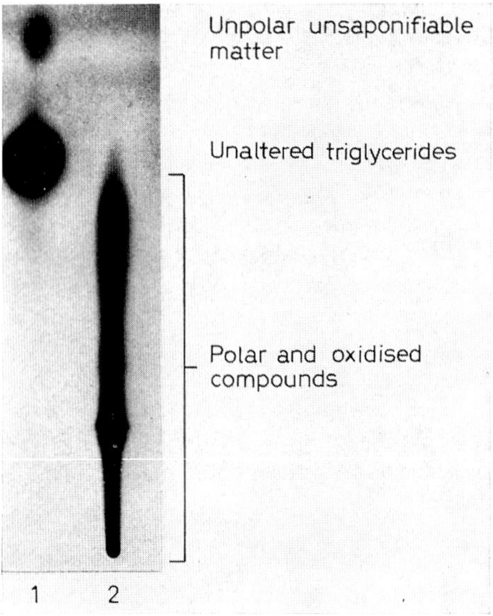

Fig. 3. Thin-layer chromatogram of the fractions obtained according to the procedure shown in figure 2. 1 = Fraction 1, unpolar compounds; 2 = fraction 2, polar compounds.

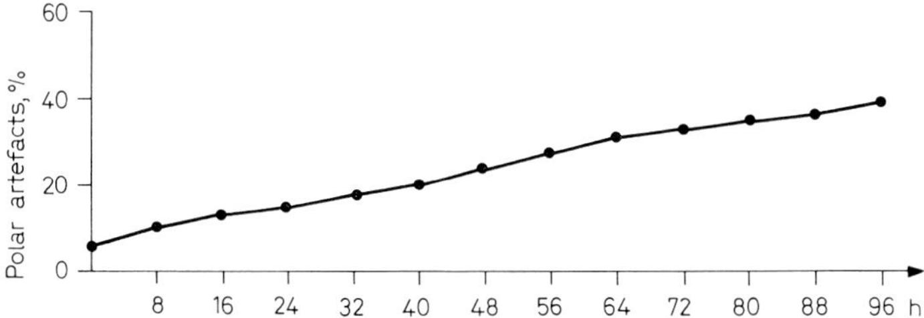

Fig. 4. Formation of polar components during deep-fat frying of pommes frites in a commercial fryer.

Table I. Preparation of fractions for long-term feeding experiments

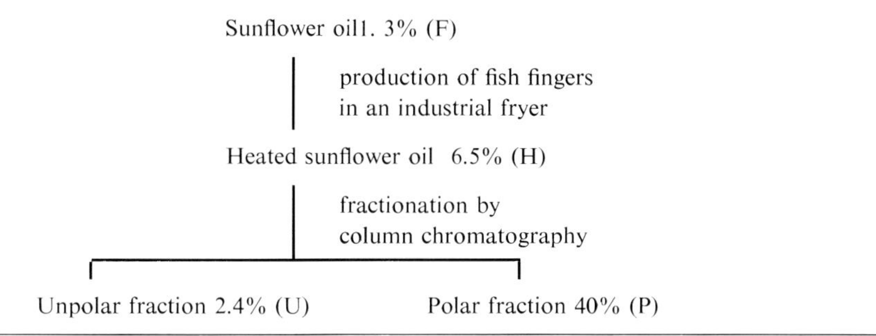

Are Heated Fats Harmful?

We started with sunflower oil which was used for the industrial production of fish-fingers. It was taken at the end of a production period when the oil was usually discarded according to the practice of this company. This oil was fractionated by column chromatography into an unpolar and a polar fraction.

We used a stainless steel column with a volume of 180 litres which was fed with 100 kg of silica gel, the solvent system was hexane and diisopropyl ether, and isopropanol to elute the polar fraction. Several tons of the used sunflower oil were fractionated according to this procedure.

Table I shows the material and the fractions we used for our long-term feeding studies: the fresh sunflower oil was used for the control group of animals, indicated by the letter F; the heated oil after production of fish fingers was given the letter H; after fractionation by means of a silica gel column we obtained an unpolar fraction U and a polar fraction P (table II).

Each fat or fat fraction was fed to rats at 20% by weight in the diet over a period of 1.5 year according to this plan. Each animal was kept in a single cage, water and diet were

Table II. Feeding plan for long-term feeding experiments

Group	Animals		Fat in diet (20% by weight)
	males	females	
F	30	30	sunflower oil
H	30	30	heated sunflower oil
U	30	30	unpolar fraction
P	30	30	polar fraction

given *ad libitum*, the room temperature was 22 °C, the relative humidity between 50 and 60%. Artificial daylight was 12 h per day. The diet was prepared once a week to avoid decomposition. The gross behaviour of the animals was checked daily. Food and water consumption, as well as body weight were determined once a week. Many of the parameters of clinical chemistry were checked after 3, 6, 12 and 18 months. After 1 year, 5 animals of each group were sacrificed for histological and other examinations.

The diet was usual feed for test animals, the test fats were given at a level of 20% by weight. Food and water uptake was the same in all groups and the diet was well accepted. The consistency of the faeces was somewhat softer as usual as was to be expected with such high doses of fat in the diet, but there was no diarrhoea.

The mortality rate was around 8% which is not unusual in feeding experiments over 1.5 year. There was no significant difference between the four test groups.

A very significant difference, however, could be observed as far as the gain in weight of the animals was concerned. Figure 5a shows the mean values of body weights of the male animals. The highest increase in weight was observed with fresh sunflower oil (F) in the diet. The unpolar fraction (U) and the heated sunflower oil (H) caused a somewhat lower increase. The group fed the polar fraction (P) exhibited growth retardation. With the female animals (fig. 5b), the differences between the various groups were not significant. This was to be expected because the weight increase of female rats is much lower than that of males, however, even in this case, the polar fraction of the heated fats (P) showed the lowest contribution to the increase in body weight.

Table III summarizes the investigations in clinical chemistry and haematology. Within this report, it would be absolutely impossible to discuss the background of all these investigations, and to report the results of these analyses. Furthermore, this is not necessary because in most cases we found

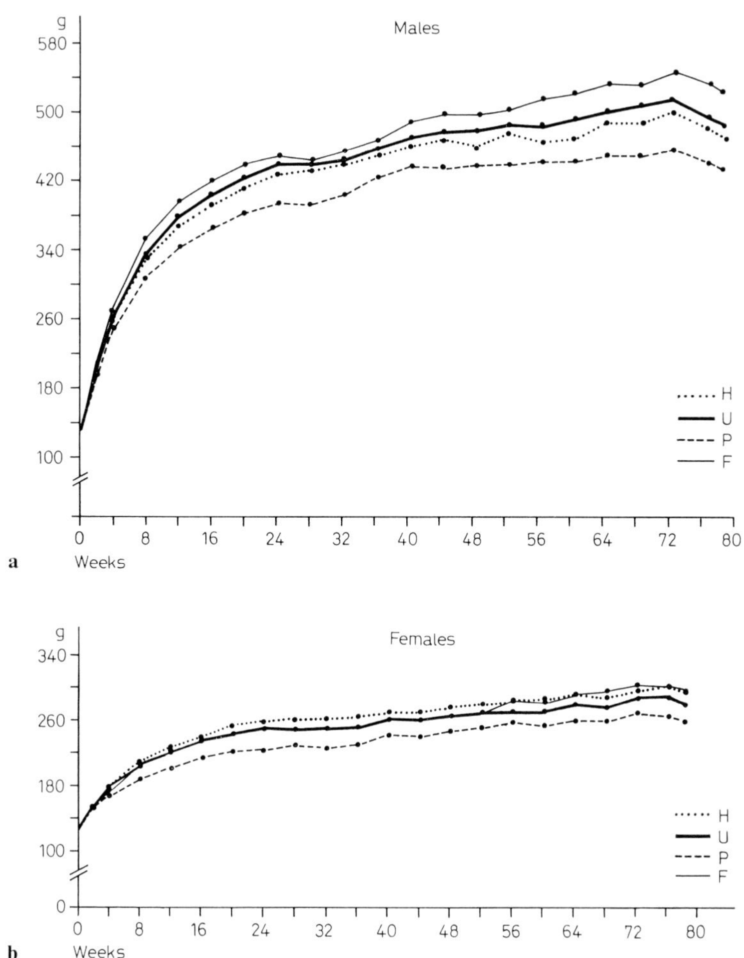

Fig. 5. Weight gain of the male (a) and female (b) animals after feeding with fractions F, H, U and P.

neither deviations from the normal values nor differences between the four types of diet fed to the animals. Therefore I would like to point out to those analyses that showed significant differences between the groups. They are indicated by an asterisk in table III.

Serum glutamic pyruvic transaminase (fig. 6) values were significantly higher after feeding the polar fraction of the heated sunflower oil. Male and

Table III. Analyses carried out

Serum	*Serum enzyme activities*
Glucose	Glutamic pyruvic transaminase*
Triglycerides	Glutamic oxaloacetic transaminase*
Cholesterol	Lactic dehydrogenase
Protein	Alkaline phosphatase
Albumin	Creatine phosphokinase
α-Globulin	Prothrombin clotting time
β-Globulin	
γ-Globulin	
Urea	*Urine*
Uric acid	Sodium/potassium
Creatinine	Sediment
Bilirubin	
Ions	*Haematology*
Sodium	Erythrocytes
Potassium	Haemoglobin
Calcium	Haematocrit value
Iron	Leucocytes
Chloride	Reticulocytes
Phosphate	

* Significant difference between groups.

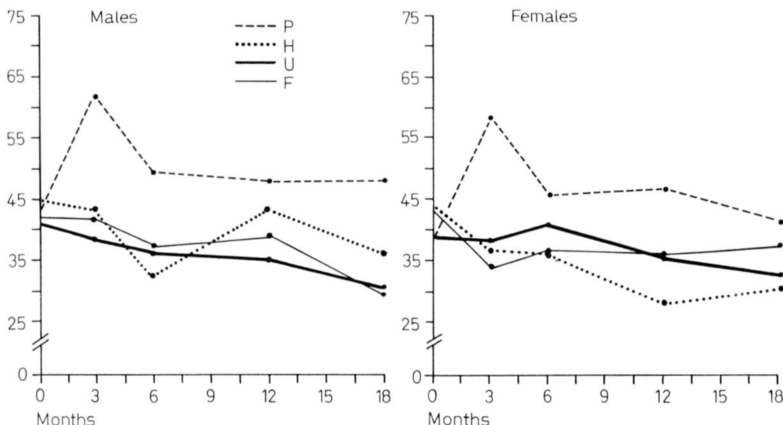

Fig. 6. Serum glutamic pyruvic transaminase values in male and female animals after feeding with fractions F, H, U, P.

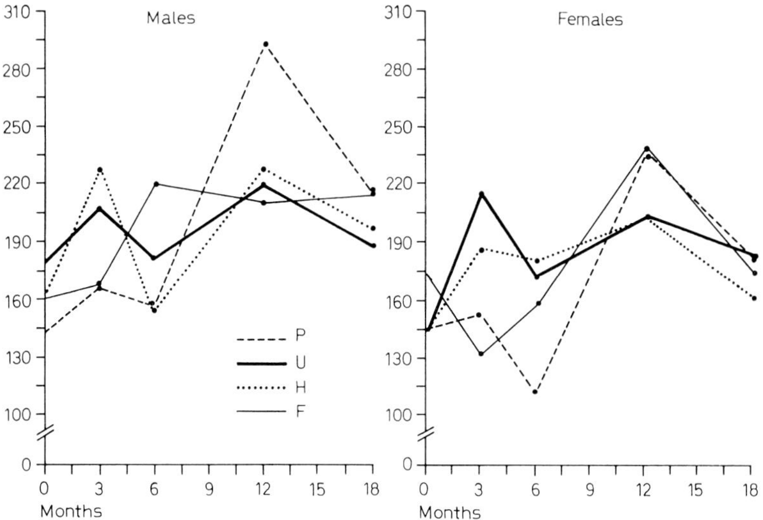

Fig. 7. Serum glutamic oxaloacetic transaminase in male and female animals after feeding with fractions F, H, U, P.

female animals showed the same effect. High serum glutamic pyruvic transaminase values are caused by increased membrane permeability and the release of enzymes of liver cells into the serum. Such values indicate some kind of a liver damage. In our case, this occurred only after feeding the polar fraction of the heated sunflower oil which had the highest content of oxidised material.

Serum glutamic oxaloacetic transaminase (fig. 7) was increased only in one instance. The mean value in the sera of male animals was rather high after 12 months. Again this occurred after feeding the polar fraction of the heated sunflower oil. These two examples are given here in detail with good reason. In all other cases of the many clinical analyses, no significant deviations were observed.

These long-term feeding experiments started with 240 rats in eight groups. 23 animals died during the experiment, and after 1.5 year, all the remaining animals were sacrificed and underwent histological and other investigations. For example, the weights of several organs, like heart, liver, kidney, adrenal gland and brain were determined. Again I would like to pick out only those two results that deviated from normal values.

The average weight of the liver (fig. 8) was significantly higher in those

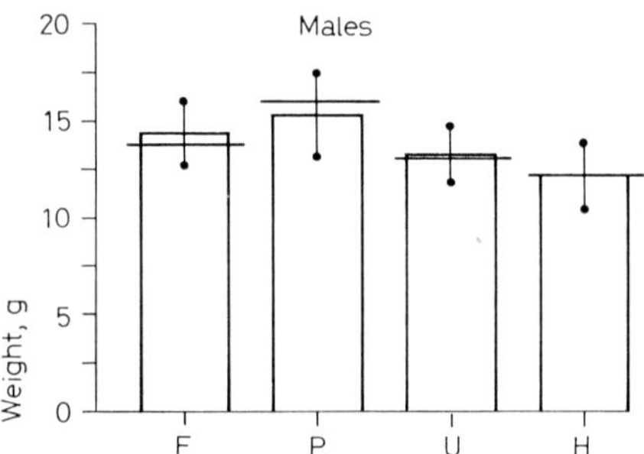

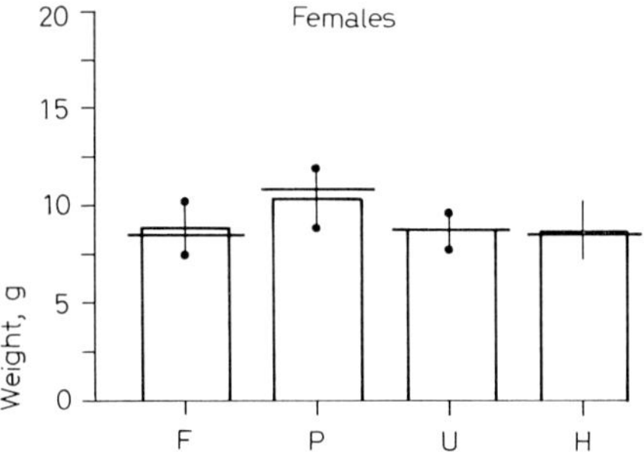

Fig. 8. Mean values of the liver weight of male and female animals after feeding with fractions F, H, U, P.

animals that received the polar fraction of heated sunflower oil. Obviously, this fraction, which had the highest concentration of oxidised material, affected metabolic processes in the liver more than the other three fat diets. A similar observation was made with the weight of the kidneys. Again, after feeding the polar fraction their weight was increased.

The histological investigations of tissues of many organs of the test animals did not show any results that are worth being mentioned here. Only the usual symptoms of the process of ageing were observed.

Conclusions

The results of these long-term feeding experiments are summarised in table IV. Only the polar fraction isolated from heated sunflower oil showed noticeable effects. It has to be mentioned, however, that all these deviations

Table IV. Effects observed after feeding with the polar fraction (P)

Growth retardation
Increased serum glutamic pyruvic transaminase activity
Increased serum glutamic oxaloacetic transaminase activity
Increased weight of liver and kidneys

Composition of the polar fraction
∞ 90% total polar artefacts
∞ 40% polymeric triglycerides

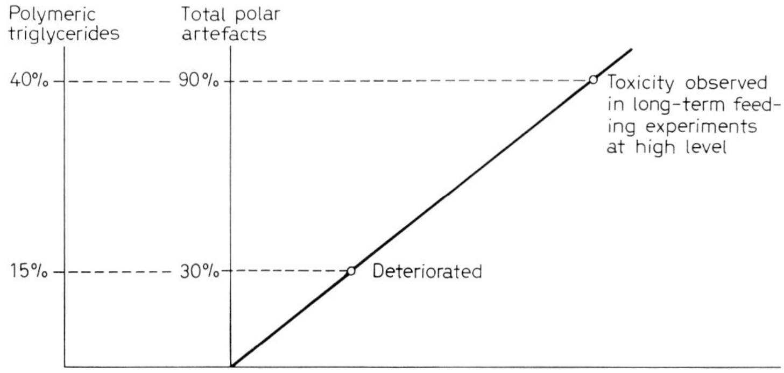

Fig. 9. Composition of heated (deteriorated) oils in correlation with observed effects, e.g. deterioration and first symptoms of toxicity.

must not be taken as being very serious. Especially in connection with the histological investigations, which did not show any severe defects, one may assume that the animals would have recovered from all these malfunctions if the diet had been changed to a normal one.

We are now in the position to complete figure 1 with additional information (fig. 9). In correlation with sensorial assessments, deterioration must be assumed if a heated oil contains 30% decomposition products; in the long-term feeding trials, the first symptoms of toxicity were observed with a fraction which contained 90% of decomposition products when this fraction was fed at 20% by weight in the diet over 1.5 year. The average daily uptake of this fraction by the test animals was approximately 10 g per kilogram body weight. If we take into account (1) that, according to statistics, humans consume an average of only 0.1 g frying oil per kilogram of body weight, i.e. 100 times less, and (2) that the usual oil quality is by no means as decomposed and oxidised as the polar fraction we used for this experiment then we are absolutely on the safe side with the proposal that a frying oil should be regarded as deteriorated and therefore unfit for human consumption if it contains 30% of polar artefacts.

Acknowledgements

I would like to express my thanks to Mrs. *G. Guhr*, a staff-member of our laboratory, who organized and performed the experimental work. The test animals have been fed by Dr. *Sterner* of 'Bio-Research Inc.' at Hannover.

Prof. Dr. G. Billek, Unilever Forschungsgesellschaft mbH, Behringstrasse 154, D-2000 Hamburg (FRG)

Nutr. Metab. *24* (Suppl. 1): 211–212 (1980)

Concluding Remarks

H. Peeters

Lipid and Protein Department, Institute for Medical Biology, Brussels

The influence of this meeting on the future resides in the contacts between people of different disciplines: this was indeed not a monoclonal meeting at all. When reading a paper one often feels that the author is repeating himself, whereas the listeners receive valuable data which can interact with their own experience.

The organizers selected several topics, namely a chemical topic on lipoproteins, a dietary topic on linoleic acid, a topic on epidemiology and lastly the nutritional aspects of fat processing. For the sake of a synthesis I propose to reconcentrate the entire meeting along the following ideas.

(1) Lipoproteins came first with problems of structure and emergence of lipoproteins, problems of metabolism connected with LDL and HDL.

(2) Next we had clinical studies such as the influence of alcohol and a discussion on lipoprotein lipids and diabetes.

(3) Specific biochemical studies were presented, namely corticosteroid biosynthesis and the effect of polyunsaturated fatty acids and their location in the heart muscle.

(4) The epidemiological papers were concerned with the trends in the world situation and gave specific data for Alabama and Belgium.

(5) Dietary studies were concerned with platelets, sugar intake, blood pressure, prostaglandins and fat intake in children.

(6) The technological aspects of fat processing came last.

Quite a lot is known about lipoproteins. We know the apopolypeptides onto which the phospholipids, cholesterol esters, triglycerides and cholesterol are attached. We know the amino acid sequence of these apopolypeptides. We are aware of the small genetic differences between the structure of these polypeptides in man and primates. Information about the metabolism of lipoproteins is also available. However, we still have to elucidate a number of things. By studying the lymph, for instance, a new apoprotein appeared which was described at this meeting. It exists in the free state in the plasma, and this arouses the question about the value and the significance of apoproteins and their function.

When one hears about lipoproteins, one automatically thinks that the main function of lipoproteins is to carry fat; this function is obvious. At the molecular level, however, and after a closer look into the structure and function of lipoproteins some specific physiological functions make their appearance, such as specific transport of specific phospholipids to cell membranes where the phospholipids – especially those which contain polyunsaturated fatty acids – are easily introduced into the membrane but also easily attacked by enzymes. I believe that one of the functions of lipoproteins would be to transport and to carry – in a protected way – these important phospholipid components towards the cell membrane and there to exchange these important building materials with the membrane. We feel that the function of lipoproteins is not merely to transport triglycerides through the organism no more than the only function of albumin is to transport fatty acids.

Many of the known polypeptides and phospholipids, even some cholesterol esters have specific functions in specific organs of the body and this is an important development. Lipoproteins, for instance, act at the level of the arterial wall where they are operative in thrombosis. Their functions lie deeper than the apparent quantitative transport of given lipid material through the organism.

This is where we have to look for in the future and this is what will emerge during the coming years.

Dr. H. Peeters, Lipid and Protein Department, Institute for Medical Biology, Alsembergsesteenweg 196, 1180 Brussels (Belgium)